TABLE OF CONTENTS

CHAPTER I   THE ENVIRONMENTAL IMPACT STATEMENT
             - AN ASSESSMENT OF U.S. EXPERIENCE

             1. Introduction   7
             2. The Environmental Impact State-
                 ment (EIS)   7
             3. Institutions involved in the
                 National Environmental Policy
                 Act   8
             4. Actions subject to the EIS
                 process   11
             5. How the EIS process has ope-
                 rated   12
             6. Public involvement   15
             7. Unresolved issues and defi-
                 ciencies of the EIS process   15
             8. Evaluation of experience with
                 the EIS process   20
             9. Conclusions   25

CHAPTER II   THE ROLE OF ENVIRONMENTAL GROUPS
             IN THE SITING OF MAJOR ENERGY FA-
             CILITIES - the U.S. EXPERIENCE

             1. Introduction   27
             2. Methodological considerations.   28
             3. The Bodega Head nuclear power
                 plant   35
             4. The Kaiparowits Plateau coal-
                 fired electricity-generating
                 plant   47
             5. The Midland, Michigan, nuclear
                 power plant   58
             6. Abbreviations, acronyms, and
                 short titles ...   75
             7. References   77

4

INTRODUCTION

The compatibility of environmental protection goals and energy policy goals has been studied for some time in the work programme of the OECD Environment Directorate. A report published in 1977, "Energy Production and Environment", discusses problems of siting major energy facilities and one important conclusion it draws is that environmental protection costs will be greater if energy policies are pursued without regard to environmental considerations or if post facto environmental protection is applied, since cleanup is usually more expensive than prevention.

Our knowledge of siting problems has since improved and is described in some detail in a 1979 report, "Siting of Major Energy Facilities", which concentrates on siting policies and procedures and the division of responsibility between national, regional and local authorities. It also gives comprehensive information on socio-economic impacts resulting from the siting activities and deals with land use constraints, establishment of power parks, recuperation and use of waste heat, etc. The set of conclusions which it has been possible to draw from the study has been adopted as guidelines for Member countries.

Substantive background information was collected to produce the report and its conclusions. Because of its general interest, this information is now being published in the form of studies on specific aspects of the question.

Chapter I describes the Environmental Impact Statement which has had a strong influence on environmental policy in the United States. Many OECD countries have adopted, or plan to adopt, some of its features. Its critics describe it as a time-consuming bureaucratic procedure which leads to delays of projects. Its supporters praise it as an instrument which has helped to introduce environmental considerations in the decision-making process of administrations.

Environment protection groups have had an important role in controversies over the siting issue. As the public became increasingly concerned with environmental matters, they gained experience in the diffusion of information, in motivating the public, in lobbying politicians to bring pressure on regulatory authorities and, lastly, defending their case in the courts. Chapter II provides a brief analysis of such actions in the U.S.A.

French experience differs from that of the United States. Chapter III describes the decision-making procedure in France for siting thermal power plants, as it emerges from the changes made in the 1970s in environmental protection legislation, the procedure for the "Déclaration d'Utilité Publique", and generally, the current relations between State and regional authorities.

In Chapter IV a different case is presented, that of a Federal State, to illustrate that land-use planning and siting policies and procedures vary widely among OECD countries and that they are closely related to constitutional and administrative traditions. These differences stand out more clearly by presenting the situation in a Federally structured country as opposed to a centrally governed one.

However, the difficulties encountered in the siting process can be traced to a number of factors common to most Member countries: intense industrialisation, high population density and pollution levels, concern for the quality of life, etc.

Many effects of siting are judged to be environmentally negative, but it is important to take an overall view and also look at the positive effects of siting. Chapter V describes the socio-economic impacts on local communities of the siting of nuclear power plants and the measures taken by some Member countries to reconcile ecological and energy concerns and obtain the agreement of the local population and their elected representatives to the establishment of these facilities.

The material set out in this publication was adopted by the Environment Committee of the OECD in April 1978.

CHAPTER I

THE ENVIRONMENTAL IMPACT STATEMENT-
AN ASSESSMENT OF THE UNITED STATES EXPERIENCE

1. Introduction

The National Environmental Policy Act of 1969 *
(NEPA) was intended to force the Executive agencies to
include in their decision making process a thorough
consideration of the environmental values that would
be affected by their programs.

NEPA contains "action forcing" provisions to ensure that the Federal agencies actually take environmental values into account when making program decisions. Section 101(A) of the Act directs Federal agencies to protect and restore the environment. Section
101(B) contains six environmental mandates that the
agencies must take into account in their decision making, but the most important action forcing provision
of the Act is Section 102(2)(C) which requires that
all Federal agencies prepare an Environmental Impact
Statement (EIS) whenever a major Federal action is to
be undertaken.

NEPA has been instrumental in raising environmental awareness to a new high level in the Federal agencies and while the EIS has been the most effective
means of accomplishing this, NEPA must be considered
in toto when evaluating its overall effect on agency
programs.

2. The Environmental Impact Statement

In Section 102 of the Act "Congress authorizes
and directs that, to the fullest extent possible :
(1) policies, regulations and public laws of the United
States shall be administered in accordance with the
policies set forth in this Act and (2) all agencies
of the Government shall : ...

"include in every recommandation or report on proposals for legislation and other major Federal actions

---

* 42 USC., 4321.

significantly affecting the quality of the human environment, a detailed statement by the responsible official on

  (i)   the environmental impact of the proposed action
  (ii)   any adverse environmental effects which cannot be avoided should the proposal be implemented
  (iii) alternatives to the proposed action
  (iv)   the relationship between the local short-term users of man's environment and the maintenance and enhancement of long term productivity and
  (v)   any irreversible and irretrievable commitments of resources which would be involved in the proposed action."

In addition, the responsible Federal official must seek help from other agencies that have special expertise and have the EIS reviewed by appropriate Federal, State and local agencies, and provide copies to the Council on Environmental Quality (CEQ) created under Title II of NEPA.

The responsible official Section 102(D) must also "study, develop and describe appropriate alternatives to recommended courses of action in any proposal that involves unresolved conflicts concerning alternative uses of available resources."

3. The Institutions That Are Involved in NEPA

a) The Agencies

All Federal agencies * must comply with NEPA thus forcing the agencies to implement the terms of the Act. Oversight of the agencies is maintained by CEQ, the Environmental Protection Agency (EPA) and Congress. The only exception to this requirement for preparing an EIS is EPA and there are still some unresolved issues such as whether such groups as the Office of Management and Budget (OMB) in the office of the President is subject to the provisions of NEPA.

At first, very few of the Federal agencies looked favorably on the requirements imposed on them by the new law. This was particularly true of those agencies with major development-oriented programs. The objections were manifold : a new, large, complicated

---

* Congress has enacted some laws which grant partial or complete exceptions to NEPA. For example, Public Law 93-153 of November 16, 1973 exempted the Alaskan Pipeline from the provisions of NEPA and Public Law 93-383, August 22 permits the Department of Housing and Urban Development to delegate its NEPA responsibilities (usually to the Mayor of a city) for its BLOCK Grant program.

8

procedure was imposed in addition to whatever existing procedures had to be followed in the development of programs, the unfamiliar new requirements interrupted the ordinary flow of planning and slowed development of agency programs. The need for preparation of EIS Statements with consideration of alternatives and review by a multiplicity of agencies not always sympathetic to the mission of the agency preparing the EIS were also sources of delay and friction. As will be discussed later, the initial delays were further compounded by the aggressive involvement of the courts in the process.

b) Council on Environmental Quality

The CEQ has two major roles under NEPA : (1) advisor to the President on environmental policy, and (2) overseer of the implementation of NEPA by the agencies. In this latter role it has prepared a series of guidelines designed, in general, to bring improved compliance with NEPA but which have heavily emphasized coordination of the EIS process. The first set of guidelines was issued in 1970 and the last revision was in 1973.

The CEQ guidelines were prepared to establish a basic structure for the EIS but the individual agencies were given sufficient flexibility to tailor their EIS's to their particular activities. As a result the guidelines of the CEQ have been supplemented by written procedures by many of the individual agencies.

In addition to the guidelines, CEQ has written a series of memoranda to the individual agencies, advising them on compliance, and has a staff review prepared for potentially interesting or controversial actions.

c) Environmental Protection Agency

Although EPA has resisted preparing EIS's of its own (with the exception of a few on waste treatment grants) the requirements that will be imposed on EPA for preparation of EIS or compliance with other sections of NEPA has still not yet been completely resolved.

Under Section 309 of the Clean Air Act * EPA could become the overseer of the compliance of other agencies with NEPA. Since OMB has not been willing to take on this responsibility and CEQ has not been able to, EPA could assume a monitoring role within the government. EPA has authority under Section 309 to evaluate

---

* 42 USC., para 1857 h-7 (1970).

the merits of any proposal as well as the adequacy of
the EIS. EPA is also obliged to comment on all EIS's
(not act as a passive consultant as do other commenting
agencies) and to be certain the statements are prepared by other agencies if they are required. Unsatisfactory agency EIS's are referred by EPA to CEQ.

EPA, since November 1972, is now required to evaluate both the adequacy of the draft EIS and the environmental merits of the project. A rating system for
each factor has been designed and if a project falls
into any of the unsatisfactory categories it is rejected and must be resubmitted by the agency.

d) Congress

Despite some reservations by a number of members
of Congress about the potential adverse effects of
NEPA on economic development, "NEPA has fared well in
Congress over the past four years". * It is not clear
from the legislative history that Congress was aware
of how far-reaching the effects of NEPA would be and
particularly those of the EIS.

Congress has held several oversight ** hearings
on agency compliance, effect of court decisions and
proposals for amending NEPA. On the other hand, one
of the major provisions included in NEPA that
it was believed would result in far-reaching changes
and improvements in the decision-making process was
the requirement that new legislation be accompanied by
an EIS. Without an evaluation of the environmental
impacts of new legislation before enactment, any adverse environmental effects can only be reduced, not
avoided, once the legislation is on the books. However, Congress has not chosen to use the EIS process
in its own deliberations (see below).

e) Courts

The courts have been the most aggressive of the
various groups concerned with NEPA in the enforcement

---

* National Environmental Policy Act, Frederick R. Anderson,
Jr., Federal Environmental Law, p. 238.
** For example, hearings before House Committee on Merchant
Marine and Fisheries, "Environmental Quality", (1970); Committee
on Merchant Marine and Fisheries, "Administration of NEPA" (1971):
Hearings before the Subcommittee on Fisheries and Wildlife Con-
servation of the Committee on Merchant Marine and Fisheries -
"Administration of NEPA" (1972); "NEPA Oversight" (1972);
"Federal Compliance" (1972); "NEPA Oversight" (1975) : "Work-
ship on NEPA," (February 1976).

of both the procedural and substantive aspects of the
NEPA legislation. This has taken two forms: (1) requiring adherence to a strict standard of procedural
compliance, and (2) imposing requirements that give a
wider scope to those provisions of NEPA that are vague.
A good part of this activism by the courts is the result of "public interest" lawsuits brought by very
active environmental and conservation groups. The courts
have granted such groups a standing to sue and thus
allowed the courts to rule on the adequacy of implementation of NEPA by the agencies.

As far as the EIS is concerned the courts have
been active in determining what Federal actions require
preparation of an EIS and on insisting that the agencies provide better justifications for determining
that an action is not "major", "Federal" or does not
"significantly affect the environment." In the preparation of the EIS's the courts have insisted that
there be full disclosure as required by the Act.

4. Actions Subject to the EIS Process

Almost all Federal actions are subject to NEPA
and the EIS process and these include (1) new legislative proposals, (2) program actions, and (3) individual project actions. These represent successive levels
of hierarchies of decision making each with a different responsibility and requiring EIS's with varying
contents. Very few EIS Statements have been prepared
for legislative programs although it would appear that
statements prepared at this early stage of program
development would be most valuable in bringing environmental consideration into the planning process.

Programmatic impact statements are being used
increasingly where repetitive type projects are being
conducted in limited geographic areas. A programmatic
statement eliminates the need for repeating, almost
verbatim, parts of the EIS for similar individual
projects. More importantly, it allows the responsible
official to evaluate the total environmental impact
of a series of individual actions, each of which might
be judged to have only a small environmental impact but
which, collectively, could have a major environmental
effect. Nevertheless, by far the largest number of
EIS's have been for individual projects such as siting
of power plants, construction of coal mines, highway
construction, water resource projects, and housing
projects.

The question of when a programmatic statement is
required has not yet been answered definitively. The
matter is still subject to wide interpretation by
agency administrators and no specific guidelines
applicable to most situations have been promulgated

11

by CEQ. The court cases that have been adjudicated to date have not, as yet, established general principles for guidance to administrators when programmatic statements are required.

The situation with respect to preparation of a project EIS is somewhat more definitive, although even here there are instances where there is a question about whether an EIS is required. In general, these doubts have been resolved in favor of preparing the Statements in order to avoid court challenges to the action.

5. How The EIS Process Has Operated

A number of problems have developed as the agencies experimented with ways to satisfy the requirements of the EIS under NEPA. The process was entirely new, imposed an additional responsibility on all agencies and was especially difficult for the development-oriented agencies to adjust to, and no precedents were available on how the new responsibilities were to be carried out. The additional effort and time that were required created a negative attitude in most administrators toward the new law.

a) The Time Required

Although additional time is required for the EIS process, now that the backlog of cases that existed when NEPA was enacted have largely disappeared, delays are usually caused now by the agency's failure to recognize that an EIS statement is required. Even this problem is occurring less frequently, as the agencies gain experience with NEPA requirements.

There are also legitimate reasons why a project should, at times, be delayed as a result of the EIS preparation. Some projects are quite complicated and require a thorough evaluation to determine if the benefits of the project exceed the costs. Prior to NEPA it is unlikely that such an evaluation would have been made and, while the project would have proceeded more expeditiously, it might have been better had it not been implemented at all. Under such circumstances, the delays caused by the EIS process are constructive, rather than negative.

In addition to these causes for delays, the NEPA process itself, which involves many opportunities for public participation, is almost certain to extend the period before a final decision can be reached. Public preferences for programs change, and small but vocal public interest groups have the opportunity to participate in the process and to intervene in the courts to slow down the decision-making process. Sometimes

12

these interventions are deliberate attempts to block a generally acceptable program; at other times it can result in greater deliberation and better decision making.

Delays in the EIS process generally occur in draft preparation, in the final statement after comments have been received and after the final statement has been issued. The average time to prepare a draft statement in 1975 varied from 3 months (Department of Commerce, Department of Housing and Urban Development) to 20 months (Bureau of Land Management of the Department of the Interior). The time between the draft and final EIS varied from 2.5 months (Department of Labor) to 19.8 months (Fish and Wildlife Service of the Department of the Interior).

A simple comparison of these average times does not take into account the relative complexities of the projects in different departments, or the experience of the personnel engaged in the EIS preparation. It also fails to reflect the type of management procedures used by the different agencies. Moreover, average times tend to obscure the much longer periods required for the major and more important Federal actions to satisfy the NEPA requirements for an EIS.

Delays after the preparation of the final EIS are generally caused by litigation on NEPA issues. However, in spite of the wide publicity given to many of the cases with major implications, there have been only 654 NEPA court cases in 5 1/2 years (starting January 1, 1970) of which 363 were brought on the grounds that an EIS was required, not that the EIS was unsatisfactory. Only 5% of the 6,000 draft EIS's were challenged in court.

b) Joint and Lead Agency

When more than one agency is responsible for an action requiring an EIS, two administrative approaches have been used - joint statements by the agencies, and the designation of one as a "lead agency" in which one of the affected agencies takes the responsibility for the EIS preparation. Other agencies with responsibilities for the project or program supply necessary data to the "lead agency." No rules have been established as to which method should be used, but CEQ has helped to arrange for EIS's to be prepared by lead agencies. Because each method has advantages and disadvantages * in preparing EIS's, both will continue

---

* Environmental Impact Statements, Report of the Council on Environmental Quality, March 1976, p. 36 and 37.

to be used. There is general agreement that, in either case, formal advance agreements among the affected agencies will help to avoid potential problems in the EIS preparation.

c) The Review Process

NEPA requires that the agency preparing the EIS request the views of other agencies (Federal, State, and local) with expertise in or responsibility for the program or project. There was a learning period when the comments received were less than adequate, but this situation has apparently improved as experience has been gained with the process.

The review process can actually be more burden-some to an agency than preparation of its own EIS's because of the large number involved. At least in the early period, agencies were not staffed with the personnel with the disciplines needed to either prepare or review an EIS, and this is still the case in some instances.

The 45 day limitation for reviewing the NEPA statements leaves very little time for an agency to obtain comments from field stations and to coordinate a reply. As a result most agencies have the comments prepared in the field. Some agencies require that these be reviewed by the headquarters office in Washington. The review process is particularly diffi-cult for EPA since it comments on all EIS's, a prac-tice not followed by any other agency.

d) The Costs of the EIS

It is impossible to estimate the costs of EIS preparation since the entire objective is to integrate it as closely as possible into the rest of the decision-making process. Other agencies are required by other laws to make at least a partial analysis similar to the EIS so that it is difficult to separate out the exact amount of the extra costs that the EIS process has caused. Even for agencies which have had to estab-lish separate procedures to satisfy EIS requirements, different methods have been used to estimate the cost of the new procedures. Given these limitations, the cost estimates for the different agencies in 1974 (as reported by the agencies) ranged from a low of ₡125,000 (Rural Electrification Administration) to a high of ₡31 million for the Department of Transporta-tion. * In addition to the major difference in how costs are estimated, this wide range results from both the number and complexity of statements prepared by

---

* Ibid., p 46.

14

the agencies and the number of statements they are required to review. In most agencies the costs associated with EIS compliance represented only 1% or less of their total budget.

6. Public Involvement

The two major impacts of the EIS requirement have been (1) to institute a formal mandatory process of analysis and consideration of the environmental effects of new programme decisions, based on adequate document-ation, and (2) to permit the public to obtain the necessary data so it can participate in the process. There is no consistency among agencies in the use of public comments. All agencies seek some degree of public involvement. This can take the form of sending copies of the statements to environmental organiz-ations and interested State and local agencies; in some instances, where the project is large and there is great public interest, public hearings, meetings and workshops are held. For example, a series of hearings was held both in connection with the proposed new legislation for the Outer Continental Shelf Leasing and for the construction of the Alaska pipeline.

Generally, however, unless the particular project has aroused major public interest or public hearings are provided for by the administrative procedures of an agency (e.g., the Nuclear Regulatory Agency), public participation in the EIS process is not extensive. On the other hand, where major projects are involved, new procedures are being used, or major policy questions about the intent of NEPA are raised, the public has used the courts to influence both the content of the EIS and the impact of its implementation. Public par-ticipation in these court cases usually takes the form of intervention by national environmental groups or local organizations formed to influence the direction of a particular program that is of direct interest to the area of the local group.

7. Unresolved Issues and Deficiencies of the EIS Process

a) Actions Subject to an EIS

The NEPA Act requires an EIS for all Federal actions "significantly" affecting the quality of the environment. Federal agencies have interpreted "signif-icant" in a number of ways and it is difficult to establish criteria for permitting an agency to de-termine what is, or is not, significant. Obviously, this depends more on the overall impact of the project than the type of project under consideration. A number of criteria have been used by some agencies to establish "significance" - numerie threshold effects, the number of environmental issues that are involved, and the type of effects that have to be considered.

Another issue that still remains unanswered is whether a project that has been authorized by Congress but which the EIS evaluation process shows will cause adverse environmental effects should be cancelled or merely modified to minimize environmental damage. Agencies have argued that a program authorized by Congress must be carried out even if it is impossible to modify their environmental consequences; those concerned about the adverse environmental effects believe that the project should be cancelled if the EIS analysis is negative.

b) Scope of the EIS Analysis

The NEPA Act requires that the EIS include an evaluation of the impact of Federal actions affecting the quality of the "human environment." Questions still remain to be resolved about what is to be considered a part of the "human environment." The courts, in interpreting NEPA, have generally taken a broad view and have directed that the physical, aesthetic, historic, cultural, social and economic dimensions must all be considered. In addition, both the primary effects (direct) and secondary effects (indirect) must be included.

This interpretation of the Act raises a number of difficult issues. If all of these factors are considered in depth, the EIS becomes so lengthy that it is unmanageable for decision makers to use. Moreover, each EIS requires an interpretation of what is "significant" for the program under consideration and it should be included in the statement.

c) Jurisdictional Disputes

Administrative problems have arisen, some resulting in legal suits brought by the public, over which agency has responsibility for determining the final decision to be reached on a particular aspect of the EIS. Although this issue occurs rather often, the best known dispute is that concerning water quality. In this instance the Atomic Energy Commission (AEC) and the EPA clashed over which agency had final responsibility for water quality standards. This particular conflict was resolved by new legislation. However, this type of clash between an agency preparing an EIS and another agency with a conflicting statutory mandate remains to be resolved.

16

Another important unresolved issue concerns projects or programs in which Federal involvement, though small, may often be crucial. In a State or private program in which the Federal share is only a minor part of the total effort, it is unlikely that the EIS could take into consideration "all the factors" relating to the project - an official requirement - if it evaluated only that part of the project with which it was directly concerned. This is an important issue because of the large number of State and private projects that need some form of Federal approval or permission to go ahead, however small it be. If an environmental impact statement had to prepared in all these cases, the impact of NEPA and the EIS process would be considerably extended.

d)   Levels of Analysis

The content of EIS's should include the type of information needed for informed decision making at each of the levels of government where decisions are made. Since important projects or programs are reviewed at various levels within the organization, a "hierarchy" of Environmental Impact Statements should be prepared to supply the data needed by the decision makers at each of these levels.

The EIS, mainly because of the litigation about programs brought by public-interest groups, has become overladen with detailed descriptive material, and in many instances has become so long that it cannot be used efficiently by decision makers. Most of those preparing the documents have a misconception about how comprehensive the EIS needs to be and tend to prepare a highly technical and scientific statement while omitting the careful analysis that would make the statement of greatest use. The EIS's generally need to have the quality of the analysis improved with more focus on matters of interest to those who must make the final decision on the project.

There is still uncertainty among those evaluating the EIS process as to whether the EIS should be used as the basic decision-making document for programs or projects or whether it should be confined to a factual analysis of the environmental costs and benefits. While making it the major decision-making document would ensure that environmental considerations are included in the overall decision process (and not merely prepared separately mainly to satisfy NEPA requirements), it would put an enormous burden on the EIS and might actually result in less, rather than more, consideration of the environmental issues. On balance, at least at this time, most observers have concluded that the Environmental Impact Statement should emphasize environmental issues.

e) Project or Program Statements

Most of the EIS's that have been prepared have focused their attention on individual projects rather than on new legislation or overall programs. This has raised a number of questions as to what type of EIS is most valuable. An EIS prepared for a project can be very site-specific in its analysis and also permits much greater participation by the public in the process. On the other hand, a "piecemeal" project approach allows a series of small, related projects, each of which may have only a minimal environmental impact, to have a cumulative overall impact of great importance. It is for this reason that some believe that a project EIS cannot adequately assess the environmental impact.

The debate over whether project or program EIS's best serve the objectives of NEPA is still unresolved. A program EIS would help to rationalize the decision-making process and help assure consideration of enviromental matters. A program EIS would also force environmental issues to be considered at a much earlier stage in the decision-making process and thus give greater attention to environmental matters before individual projects are selected. However, a requirement to prepare a program EIS raises a number of difficult problems. How does one identify the appropriate forum for programs which frequently overlap among different agencies? How does one determine when a program statement is needed, since it generally has no precise beginning or ending as most projects do? Can an EIS be used effectively to analyze policy issues?

To overcome these difficulties some agencies have used both program and project EIS's. The Corps of Engineers has prepared multi-project program statements for specific geographic regions and then individual EIS's for projects. The Bureau of Land Management and the Forest Service have used a "tier system" for EIS's. A program EIS is prepared that describes the national program that is being planned, a regional EIS is prepared for each geographic area, and project EIS's are written for each project.

Program impact statements are useful in bringing the upper levels of decision makers into the environmental debate, but the further removed such statements are from specific actions, the more difficult it is to have statements prepared that can be challenged by the public.

f) Role of the Administrators

The usefulness of the EIS depends on the role that top administrators are willing to assign to

the EIS process. Most of the EIS's to date have been
prepared for individual projects and this requires
the involvement of middle and lower-level administrators. Unless the top administrative levels make a special effort to become aware of the EIS analysis   so
that they are knowledgeable about environmental issues
when they propose new policies or large new programs,
the EIS will have little effect on major agency actions.
There is considerable variation among top agency administrators in how much importance is attached to the
EIS process, but many observers believe that environmental issues are now given greater importance at the
lower administrative levels.

g)   The EIS and Congress

Although Congress has been one of the most influential defenders of NEPA and the EIS and has acted to
ensure its continued use in Executive Agency decision
making, it has not relied on EIS's for legislative
purposes. Except for few rare instances (when an EIS
was prepared for other reasons), no EIS has been prepared in connection with new legislation. It appears
obvious that an EIS at this stage of policy making
could have a greater impact than at any other stage
of program or project development. No EIS is required
in connection with the introduction of new legislation
nor have EIS's been used generally in the deliberations
of Congress. At the present there is no indication
that this situation will change. No efforts have been
made to alter this situation either by Congress itself
or by outside groups who might be in a position to
force a change to occur.

h)   The EIS Review Process

Although it is difficult to quantify the effectiveness of the impact review process, it is obvious
that better reviews are being prepared than in the
early stages when the EIS process was still under development. However, it is possible to determine how
much consideration the preparing agency gives to the
reviews of the other agencies by examining how many of
the substantive comments were incorporated in the final
EIS. According to one witness in the 1975 Congressional Oversight Hearings on the Administration of the
National Environmental Policy Act, the preparing agencies have paid little attention to the reviews and
have done not more than the minimum to make the
original draft acceptable so that it could stand up
to court review, if necessary.

i)   Delays Caused by NEPA

The delays in projects and programs caused by the
need to prepare an EIS have decreased as experience

19

was gained in their preparation. Even the Federal agencies, which have frequently been unsympathetic to the entire NEPA process, do not judge the current delays to be significant. Most of the lengthy delays were caused by court suits, and the number of these has been decreasing. Moreover, CEQ and environmental groups believe that the delays have been constructive since the EIS's were useful in mitigating adverse environmental impacts of proposed projects.

8.   Evaluation of Experience with the EIS Process

The enactment of NEPA was intended to bring fundamental reforms in the Federal decision-making process, particularly on environmental matters. It has, in fact, turned out to be the cornerstone of all other environmental legislation that has been passed since NEPA was enacted. While NEPA requires much more than the mere preparation of an EIS, other parts of the Act have not as yet been widely used in making NEPA more effective. The EIS, the major action-forcing provision of NEPA, has almost completely dominated the NEPA process and is responsible for most of the benefits that have resulted from the law.

Most Federal agencies and OMB resisted the new provisions of the Act but were pressured by CEQ, the Congress and the courts into taking the law seriously and, at least, following the EIS preparation responsibility in a satisfactory manner. It is generally thought that the transition period between resistance and the inevitable acceptance of the Act occurred in 1974 when the agencies realized that a satisfactory EIS could not be prepared by someone on the administrative staff after a project had gone forward for final approval.

The early period of confusion in the preparation and reviewing of the EIS's was the result of a number of factors, some of which are discussed in Section 5. One of the chief difficulties was that projects in progress prior to enactment of NEPA were not excluded from its provisions. Some of these projects were nearly completed and an enormous amount of effort had to go into preparing an EIS aimed at making certain that the project, which may have been underway for four or five years and nearly completed, was not stopped. This was unfortunate for several reasons. The initial adverse attitude of the agencies toward accepting NEPA was reinforced because of the pro forma way in which the EIS's had to be prepared for on-going projects. It also developed a view in the agencies that the EIS process would really not influence the new projects subject, to NEPA. Finally, it was during the learning period in which the agencies were adjusting to the provisions

of new major legislation with an entirely new legislative intent, that the least useful and most difficult
EIS's had to be prepared. The agencies' negative attitudes toward NEPA were further aggravated when a number of early EIS's were litigated in the courts and
the courts very often ruled in favor of the plaintiffs
and not the agencies.

In 1972 the General Accounting Office (GAO) investigated how well the intent of the NEPA Act had been
carried out. The report concluded that the EIS's were
"not being implemented in a uniform and systematic
manner." Improvements in preparing EIS's were said to
be needed in the following areas :

(1) Using the EIS's for decision making
(2) Defining the actions that require statements
    and establishing what environmental concerns
    were to be considered in the EIS
(3) Improving public participation in the EIS
    process
(4) Making certain that the views of the Federal,
    State, and local agencies were taken into
    account.

By 1976 more than 60 agencies had adopted forma-
lized regulations to ensure that there be compliance
with the EIS process and NEPA. In spite of this apparent compliance,the testimony given at the 1975 Oversight Hearings concluded that there have been "occa-
sions when agencies have not been particularly responsive"to the CEQ coordinating authority.

Although many critics believe most of the apparent benefits of the EIS are more procedural than substantive, there still remain difficulties with the EIS
process itself. The EIS's have been of inconsistent
quality, shallow in critical analysis of impacts, and
have caused strains on the Federal decision-making
process. These strains include a slowdown in decision
making, creation of friction among those responsible
for implementing conflicting policy objectives, difficulty in coordinating NEPA with other laws, and
problems of budgeting time and money for the EIS process.

One important aspect of the EIS process, not often
considered in the evaluation of the process, is its
impact on international policy. NEPA and the EIS's
have affected the international community in several
ways. Some U.S. agencies have used the process in
evaluation of their international programs. The United
States also has prepared EIS's in negotiating some
international agreements (Ocean Dumping, Endangered
Species, etc.) and in connection with internal U.S.
projects that affect neighboring countries - Canada
and Mexico. More importantly,some countries have

adopted requirements for the preparation of statements similar to the EIS and a number of other countries are actively engaged in determining whether such an EIS requirement would be an overall benefit to their social and economic systems.

How well the major objective of NEPA - to influence the Federal decision-making process on matters of the environment - has worked is difficult to measure. It is impossible to determine if the same decisions that have been made would have been reached in the absence of NEPA and the EIS requirement. The Corps of Engineers, as early as 1973, reported that the EIS process was responsible for 24 projects being dropped, 44 projects delayed and 197 projects modified. If it could be firmly established that the EIS had been the cause of such changes, there could be little disagreement about the importance of its impact on the decision-making process. However, there is no way to know for certain which of these projects would have been altered in the absence of the NEPA and the EIS process.

In spite of these uncertainties, the Corps of Engineers indicated, as recently as the 1975 Congressional Oversight Hearings, that NEPA had some positive effects on their agency activities. These included an overall improvement in decisions, predesign modification in projects, a better ability to consider alternatives, better coordination with the public and an improved balancing of environmental and economic development considerations.

In general, by 1976, most agencies reported that the need to prepare an EIS had been an important aid in planning and decision making, and this has been true at a number of levels in the decision-making process. Even those who are still critical of the importance of the EIS process in bringing about improvements in the environment admit that NEPA, through the EIS process, has had a salutary influence on the bureaucracy and has changed the perspectives of a number of agencies concerning the importance of the environment. The major effects are reported to be :

    (1)   A reordering of national priorities so the environmental values are part of the decision-making process.

    (2)   Opening up the decision-making process to public view.

    (3)   Forcing agencies to explain their decisions and to take criticism of them into account.

    (4)   Providing a means for forcing the agencies to give consideration to the views of those critical of their actions by bringing the issue to judicial test.

(5) Making certain that consideration was given to alternatives to a proposed action and that the interrelationships of such actions were examined.
(6) Mobilizing environmental experts able to analyze potential environmental effects of new projects.
(7) Improving coordination among agencies.
(8) Stimulating new environmental research that was needed.

Many of the changes brought about by the requirement for preparation of an EIS have been subtle but important. The unseen effects of the EIS preparation on the attitudes of the agency personnel, the development of adequate environmental staffs and assessment methodologies have resulted in improved implementation of the EIS's. The EIS process has forced interagency discussions to be held and these have been used as a lever in forcing agencies to consider the adverse environmental effects called to their attention by those agencies responsible for preventing further environmental damage in a particular area. The criticism of citizens who object to particular projects and programs because of the impacts on their surroundings (even though they may be of benefit to other segments of society) have forced the agencies to modify programs so as to avoid lengthy and time-consuming litigation.

However, many critics remain sceptical as to the positive effects of NEPA. They argue that the apparent acceptance by the agencies of the new environmental policies is "more cosmetic than genuine" and that the reduced friction and smaller number of court cases that have occurred over time is the result of learning to meet the statutory requirements of the Act while meeting only its minimal requirements for improving the environment. The agencies are said to have learned how to prepare EIS's that are acceptable to the courts although they do not result in improved environmental values.

Evidence of the effectiveness of the EIS and NEPA in reducing pollution is mixed. The National Wildlife Foundation has attempted to measure quantitatively a quality index for a number of different environmental categories and concluded that the results "have been bitterly disappointing" and that there was a significant deterioration in environmental quality in every major category between 1970 and 1974.

There is general agreement that the goal of integrating environmental policy into all actions of the Federal government had not yet been achieved even as late as early 1976. Some agencies claim to have used the EIS and the NEPA process as a mechanism for planning, for consideration of alternatives and for decision making, but the many critics of agency administration of NEPA assert that the EIS's are not really taken seriously in decision making.

Much remains to be done if the EIS process is to be used as a major decision-making document. The court requirement that agencies include in the EIS alternatives to the proposed action that are not within the purview of the agency, has contributed greatly to the length and complexity of the EIS. As a result, the increasing length and descriptive nature of the material has made it more difficult to use the EIS as a decision document. In addition, some critics believe that the EIS's have ignored the critical issues and facts relevant to the needs of those making the decisions. The EIS is of little real use if it does not address the real viable alternatives. The purpose should not be to avoid law suits but to assist in the best decision making.

On the other hand, attempts to use the EIS as a document for trading off all of the factors involved in project or program decisions may be doomed to failure. It would almost certainly load the EIS with more responsibilities than it can hope to bear. Moreover, enlarging the scope as far as has already been done has made it almost impossible for the conservation and other public interest groups, who have been the external monitors of the EIS, to use the document effectively since its growing length and complexity (and reduced analytical attention) require experts, not available to such organizations, to review and criticize the document.

A compromise position between these two views of the EIS process that seems to be emerging at present is one which de-emphasizes the procedural aspects and shifts from a project-oriented to a program-oriented EIS. The focus on the EIS to date has been largely on individual projects and the shortcomings of these types of statements are described above. By requiring an EIS on programs, which is at least one tier above the level of projects, the Environmental Impact Statements can become more heavily involved in the early planning phase and avoid the cumulative, but individually small, adverse environmental effects that are not obvious when an individual project EIS is considered. Despite the difficulties of preparing an EIS at the program level, and the scepticism reflected in some of the court decisions about the usefulness of program EIS's compared to project EIS's, program EIS's would help to redirect the misplaced emphasis on procedural requirements of NEPA to its real purpose of a "thorough analysis of environmental impacts and options to the proposed action."

24

In spite of nearly six years of experience with NEPA and the EIS, some basic questions have not been completely answered. These include (1) a definition of the scope of the environment that NEPA is to preserve. Is the environment the natural ecological system or does it include man and his systems ? (2) how can the lack of commitment to NEPA of some of the agencies be changed so that the process can be used more effective- ly ? and (3) what means can be used to assure fur- ther improvement in environmental quality, however it is defined ?

9. Conclusions

Most observers have concluded that, despite ob- vious needs for continued improvement in the EIS pro- cess, some significant results have been accomplished by NEPA, mainly through the action-forcing EIS process. However, there is also general agreement that more needs to be done than has been done so far. With the public awareness of the environment, its active parti- cipation in the EIS process, and the general accep- tance that environmental enhancement is in the public interest, there is a favorable climate for further improvement. Some of the suggestions that have been made to achieve better implementation of NEPA through the EIS process are :

- Preparation of programatic impact statements on overall policies so that the environmental impact is taken into consideration at the highest levels of government.
- New guidelines by CEQ to improve the EIS pro- cess so as to reduce the quantity of material and markedly increase its quality.
- Better funding, coordination and considera- tion of research in those aspects of environ- mental sciences for which the data base is now so inadequate that it is not possible to prepare a satisfactory EIS.

Continued oversight of the EIS process by CEQ, EPA, Congress and the courts can be expected and this should result in better implementation of NEPA through preparation of improved EIS's. Environmental matters are still of great public interest and this should guarantee that the EIS receives the attention that will result in further improvement in its preparation and use. As experience is gained with EIS preparation and the number of unresolved issues about them decrea- ses, a better environment will be created than would otherwise have resulted . No definitive study has been made to determine if the additional costs (of EIS preparation and changes in project and program plans made for the purpose of environmental

enhancement) have or have not exceeded the benefits of
the changes made as a result of Environmental Impact
Statement analysis. Until studies of this type have
been completed, it is impossible to "prove" that the
EIS has been beneficial. There is no doubt that the
EIS process has resulted in an improved environment.
However, measuring the costs and benefits is still too
imprecise an art, in this field, to be useful in evaluating the overall value of the process. Better scientific information may help overcome this difficulty,
but it will be a long time before cost/benefit analysis
will be sufficiently accurate to be the key element
in the decision-making on a new program or project.
As a result, other tests will have to be used to
balance environmental and other effects of new projects,
and the EIS process is a useful tool to assess this
balance in a systematic way.

CHAPTER II

THE ROLE OF ENVIRONMENTAL GROUPS
IN THE SITING OF MAJOR ENERGY FACILITIES:
THE UNITED STATES EXPERIENCE

1. Introduction

This chapter describes the role played by private
citizens and citizens' groups in public policy decision making in the United States, specifically the
role of environmentalist public-interest groups in the
siting and design of energy facilities. The format is
that of the case-study method, a method chosen because
it gives insight into the phenomenon by revealing
properties common to most instances of involvement by
environmental groups in energy facility siting issues.

Each type of energy-supply system has unique environmental aspects as well as social, economic, political, and legal ramifications. Nonetheless, it is
hoped that the studies reveal similarities in the role,
function, motives, and strategies of environmental
groups.

The chapter touches upon several components of
energy-supply systems (extraction, conversion, and
transport), as well as upon power-generating facilities based on different sources of fuel (fossil fuel,
nuclear fission, etc). We attempt to analyse the involvement of citizens in public policy decision making
at all levels of government.

The case studies epitomise various aspects of the
roles, functions, motivations, and importance of environmental groups in determining public policy. The
evaluation reflects the dynamic process of participation within a fluctuating legal framework. The studies
describe patterns of organisation and participation
by the public that have evolved since 1960. The studies provide typical examples of the various ways in
which citizens can intervene and the strategies they
use to do so. The case studies demonstrate the usefulness of the courts in establishing environmental
values.

Attempts are made to analyse not only the goals
and results of citizen intervention, but the under-
lying reasons for the goals and the reasons for the

27

results, as well. Most importantly, it is attempted to assess the efficacy of these participatory processes, while avoiding the question of majority rule and minority rights in the American democratic system.

The cases are arranged in chronological order, with the National Environmental Policy Act of 1969 (NEPA) assuming a central point of reference. NEPA is generally considered the most important piece of national legislation on the environment in the United States. It has played a critical role in creating, defining, and clarifying channels for public participation. Therefore NEPA was used as the primary reference point : the Bodega Head controversy took place before NEPA was enacted; while the Kaiparowits, and Midland cases occurred after (and in part because) NEPA was enacted.

Each case study has (1) a chronology of significant events in the case;(2) a narrative description of the events; (3) an analysis of the role played by environmental groups; and (4) a list of references cited. (5) A short bibliography or (6) an annex is appended if necessary.

2.   Comments on the Methodology

a)   Identifying Environmentalists

In preparing the case studies that follow, it was difficult to distinguish the role of environmentalists in the public decision-making process from that of other political activists. This was so because no single c r i t e r i o n identifies an environmentalist. A public opinion poll, commissioned by the United States Senate's Sub-committee on Intergovernmental Relations, found that 75 per cent of the sampled population were members of one kind of social group or another, although only 10 per cent of them could be defined as active (O'Riordan, 1976).

The nature of the energy problem itself, an underpinning of modern industrial nations, brings together a broad range of interests. Moreover the nature of each issue determines how the public will become involved. Each public participant has a unique combination of motivations, goals and functions that, in turn, determine his or her organisation, strategies, and action. When one considers the breadth of environmental issues and concerns, together with the many ways of organising and the growth of the environmental movement, it is obvious that categorisation is difficult.

Most participants in conservation issues are middle-class, urban, and professional people; however,

28

citizens' groups now draw their members from a far broader socio-economic range than previously (Sewell and O'Riordan, 1976). In addition to the broadly based growth in citizen involvement, especially in environmental issues, the goals and techniques of organised environmental groups have shifted from hiking and similar limited interests to educating the public and influencing public policy.

b) Identifying the Goals and Motives of Environmentalists

The specific goals and motivations of citizen participation cannot be easily characterised. Above all the "participatory process" reflects those desires of the public that are not effectively represented in the political forum. Public participants strive to balance decision making and to establish effective outlets for information and the expression of views. Public participation is based on protecting minority views; seeking economic, political and social equity; thoroughly examining public needs; establishing responsibility and responsiveness to institutions and institutional arrangements, providing for due process; promoting public interests; and assuring that sound and wise decisions are made (Wengert, 1976).

Citizens become involved in public-policy issues for reasons that depend upon their own various interests and points of view, the political and economic context, and existing institutions. "Demands for more public participation may be motivated by a desire to alter the power structure and thus weaken the establishment" (Wengert, 1976), or they may simply seek to ensure the input of better information and more service from public officials.

The disillusionment and disappointment of citizens in public policy making on issues of environmental concern, paralleled by a growing distrust of official decision making on highly technical questions such as the nuclear power issue, has widened the so-called "credibility gap". Over all, the problems have opened the regulatory agencies and the government's entire policy-making process to criticism and scrutiny by the public.

The planning and construction of energy facilities employ resources in direct competition with recreational, aesthetic, or other uses and values. The desire to preserve such environmental amenities may lead citizens to become actively involved. Often, concerns other than potential environmental degradation lead citizens to participate in decisions on the siting of energy facilities. In the case of nuclear power plants, for instance, many people, especially those who intervene,

29

are concerned - even afraid - that catastrophic acci-
dents, the release of radioactive compounds (especially
over the long term), and technological error, could
occur. Moreover, citizens may mobilise to oppose a
nuclear power plant in their neighbourhood because of
its effects on environmental or aesthetic values
(Ebbin and Kasper, 1974).

c) Organisation of the Environmental Groups

The concerns of citizens are converted into effective political and social responses primarily through
public-interest organisations (Messing and Rosen, 1972).
But the term "public interest organisations" encompasses
a broad spectrum of organised groupings that, at times,
emphasize environmental concerns. All vary in their
roles, functions, choices of issues, and strategies.
Among the list are the old-line conservation, environmental public law, and other special-interest organisations, all of which may organise in coalitions, informal and ad hoc groupings. In addition to these categories of organisation, a public-interest group that
focuses on other than environmental concerns may also
have over-lapping or contiguous interests, as exempli-
fied in alliances with labour groups and consumer activists (e.g. the United Auto Workers, alliance with
the environmentalists in the Midland, Michigan nuclear
power plant case study).

d) Institutional Arrangements

The complex nature of energy issues and the uncertainty of the short- and long-term effects of a variety
of possible decisions, coupled with existing institutional arrangements, limit the ways in which environmental groups may effectively intervene. Moreover,
measuring the effectiveness of any one participatory
action is difficult. Each effort at participation is
an attempt to influence a spectrum of decisions by
governmental bodies, and often at more than one
level of government. The complexity of processes like
litigation, or licensing and permit procedures via the
regulatory agencies, requiresvarious degrees and types
of input, from grassroots activism to sophisticated
legal, economic, scientific, and technological intervention in decision-making procedures. The situation
is further complicated by interactions among special
interest groups, several regulatory agencies, utilities, and industry.

e) The Function of Environmental Groups

Environmental public-interest groups have two
basic tasks in energy siting and design issues: educating the public, and intervening in public policy
making and litigation. The public education aspect

30

involves disseminating information, gaining public attention, and organising, maintaining, and building the constituency of a group. Political and legal intervention is achieved by various strategies that require more sophisticated and specialised expertise.

Emphasis of function is one determinant of a group's organisation that, in turn, influences the choice of strategies and methods of intervention. Another important element is the scope of issues to be addressed. A group may be organised around one specific issue, or may be based upon ideology, extending its scope to numerous environmental concerns. Generally, issue-oriented action groups are organised so as to maintain a well-defined collective self-interest (O'Riordan, 1976). This structure requires that personal interest be high and that strategies be well supported. In contrast, ideology-based environmental groups may lack focus, definition of goals, or well-supported strategies.

On the whole, citizens become involved in public policy issues most effectively when they deal with small-scale specific issues, especially at the lower levels of government, where their efforts and representation are more concentrated. As more people become involved and as more special interests are represented, the problems become more complex, the geographical coverage widens, and the familiarity of members with specific issues diminishes (Sewell and O'Riordan, 1976).

f) Public Education

The public education aspect of environmental groups involves the grassroots activism of many organisations. This is also the sole function of some organisations. Diverse resources are available for organising, informing, and publicising environmental issues. The choice of technique depends primarily upon the intended audience. Public-service announcements, feature articles in newspapers, and articles in popular magazines and books are outlets for information and the expression of views aimed at a broad cross-section of the public. Specialised or topical news reports and technical articles are means of reaching more select audiences with higher levels of expertise. Writing letters to political officials to express particular environmental concerns is a technique aimed at a very specific audience. Another technique, effective when it coincides with a political election, is the use of "voting charts" (tabulations of the voting records of candidates on specific environmental issues). Political demonstrations and public rallies are important mass media and educational tools. Workshops, conferences, and task forces, organised around specific issues, are commonly used as more sophisticated, public-educating techniques.

g)   Procedure for Intervention

The environmental movement, like every other great
social movement in the United States, eventually turned
to the legislatures and the courts to define the roles,
rules, procedures, and social norms for effective participation by the public (Sive, 1973). Three comple-
mentary federal laws guarantee citizens access to and
involvement in political and government institutions.
They are :

(1)   The Freedom of Information Act (FOI),
(2)   The National Environmental Policy Act of
      1969 (NEPA), and
(3)   The Administrative Procedure Acts (APA).

These laws strengthen specific environmental law,
allow access to information, and open public meetings
to observance - and at times participation - by citizens. NEPA is the most important of the three laws
because it is broadly based and all-encompassing. It
is considered to be the most comprehensive legislative
statement ever made in the United States on the environment (Anderson, 1973).

h)   NEPA and the Human Environment

The enactment of NEPA both reflected and contributed to a changing attitude toward the increasingly
threatened human environment. It defined a point of
entry through which citizens could become involved in
the setting of public policy. Therefore, NEPA, as a
positive public policy, opened new avenues to decision-
making procedures, although its ultimate effectiveness
had not been fully anticipated by the legislators.

The broad and all-encompassing scope of NEPA
threatens its effectiveness. However, a key "action-
forcing" provision - Section 102(2) (c) - ensures that
agencies of the federal government will consider the
environmental consequences of their planning and decision making. NEPA is far more explicit in its require-
m e n t s for changes in organisation than in its suggestions for substantive changes in policy. Before
NEPA, there were no formal guidelines or principles
for public participation. The changes NEPA mandated
include the environmental impact statement (EIS) requirement, the use of interdisciplinary decision making, the inclusion of "unquantified environmental...
values" in a benefit-cost context, the specification
of alternatives, and the creation of a new organisation in the Executive Office of the President, the
Council on Environmental Quality (CEQ), to oversee
the progress (Culhane, 1974).

i) Regulatory Agencies

Since the inception of the administrative process, the responsibility of governing the environment has shifted to administrative, or regulatory, agencies. Today, regulatory agencies are characterised as overly responsive to the various industries they were supposedly created to regulate (Culhane, 1974).

The "action-forcing" feature of NEPA is the requirement that regulatory agencies prepare an EIS before every major action that significantly affects the quality of the human environment* . This was the basic stimulus for procedural changes in agencies, including the institution of public participation programmes.

In general, the public participation programmes assume three basic forms (Culhane, 1974).

The first form is written comment on agencies' EIS's, a type of participation regarded as only an entry into the full participatory process. The second form is the participation of citizens in public meetings and hearings, but here the public is generally requested only to indicate its dislikes about a pre-selected plan or policy. This has implications on the process of decision making and its outcome. The third form of participation, considered by all parties to be the most useful avenue for achieving goals, includes informal contacts, typically with one or more administrators, and a restricted participation of agency clientele.

Overall, NEPA and the EIS process it instituted provide a basis for exerting effective pressure on agencies; public participation programmes are a response to accommodate these external pressures. NEPA provides the means to force environmental accountability on agencies, although it may not force agencies to become optimising decision makers (Friesema and Culhane, 1976).

In the case studies, the efficacy of several agencies' public participation programmes are examined, with examples both before and after NEPA. It is difficult to measure the relative extent and effectiveness of public participation programmes because there are considerable differences among agencies, and even among different levels of one agency.

---

* Two EIS's are prepared on every proposed action : (1) a "draft EIS", prepared by the regulatory agency and circulated to other relevant agencies and to the public, and (2) a "final EIS", which incorporates and addresses the comments made by other governmental agencies and the public on the draft EIS.

j) The Legal Interpretation of NEPA

In the United States, the judiciary is independent
of Congress and the state legislature. Therefore, it
is capable of interpreting statutory legislation, and
also of pointing out inconsistencies, failings and
loopholes in legislation, thereby prodding legislators
to review their policies. One trend of the courts at
NEPA's inception was to tighten the review of agency
decision making. NEPA's reform-minded provisions contributed to the courts' efforts in a similar fashion.
Thus the courts found a surrogate review mechanism in
the "Section 102 process" (Anderson, 1973).

The favourable interpretation of NEPA by the sympathetic court of the early 1970s, under the leadership of Associate Justice William O. Douglas, contributed to the environmentalists' high regard for adver-
sary proceedings and the role of the courts as effi-
cacious mechanisms for public participation (Sive,
1973):

> "Important environmental controversies not basic
> or general enough to be susceptible of express
> legislative determination should be aired and en-
> lightened by a litigating process. That process
> has as its principal features examination and
> cross-examination and reasoned exclusion of what
> is irrelevant from the basis of determination...
> it is believed that in any environmental contro-
> versy involving the weighing of conflicting val-
> ues, the weigher should be a court, a generalist,
> rather than an administrative agency whose out-
> look is organically developmental and provincial...
> Claims and counterclaims somehow achieve legiti-
> macy and importance when made within the bounds
> of a legal proceeding. Whatever the explanation,
> an environmental law-suit can be and has frequent-
> ly been an effective political instrument."

In initial interpretations of NEPA, the courts
were more willing to review compliance with the procedural requirements than to enforce substantive positions upon the regulatory agencies. However, more
recent litigation demonstrates that the courts are
willing to enforce substantive rights, as illustrated
in the court opinion of "Sierra Club versus Froehlke"
(5 FRC 1065, 16th February, 1973):

> "This court is fully aware that many cases have
> held that NEPA does not create 'substantive'
> rights, but rather creates only 'procedural'
> obligations. To the extent that NEPA does not
> articulate acceptable levels of air, water, heat
> and noise pollution, this is an accurate posi-
> tion... But to the extent that it would allow
> the agencies merely to disclose the likely harm

34

without reflecting a substantive effort to prevent or minimise environmental harm, it is not an
accurate position of the role of the courts under
NEPA."

The perpetual litigation due (from the environmentalists' viewpoint) to the remarkably successful
use of the courts within a moving legal framework, is
perhaps the greatest problem in evaluation under the
case-study approach.

3.   The Bodega Head Nuclear Power Plant

   a)   Chronology

   A chronology of the major events in the Bodega
Head Case follows :

1954   The Atomic Energy Act is passed by Congress;
          it establishes the authority of the Atomic
          Energy Commission (AEC) to control and promote
          the peaceful use of atomic energy and assures
          private electric companies ("utilities") that
          the government will not compete in the sale
          of electrical power produced from nuclear re-
          actors; it also offers the utilities a suffi-
          cient research and development budget.

1955   The National Park Service (NPS) recommends
          Bodega Head as a potential site for preserva-
          tion; concurrently, the California Division of
          Beaches and Parks completes its study and re-
          commends the site for a state park.

1956   The University of California at Berkeley (UCB)
          announces that Bodega Head is the most suit-
          able site for its marine biological laboratory.

1957   Both the Division of Parks and UCB stop nego-
          tiating for the site.

1958   - 23rd May : Pacific Gas and Electric Company
          (PG & E) announces that preliminary negotia-
          tions are underway to establish a power plant
          of unspecified design at Bodega Head.

          - July : The Division of Parks announces that
          it is negotiating with PG & E to purchase land
          not used by the utility; PG & E files condem-
          nation proceedings to acquire the Gaffney and
          Stroh properties.

1959   - October : The site of the power plant is
          shifted to "get off fault".

          - 19th November : The Sonoma County Board of
          Zoning Adjustments approves powerlines over
          Doran Park.

35

- 24th November : Without public hearings, and ignoring a petition bearing more than 1,300 names against the powerlines, the Board of Supervisors grants use permits for powerlines.

1960   - 25th January : The original submittal from the Harbor Commission to the Board of Supervisors details harbour development and proposed roadway; private hearings are held to determine the fate of the project.

- 4th February : In identical resolutions, the Board of Zoning Adjustments and the Planning Commission recommend to the Board of Supervisors that it grant the land-use permit for construction of the powerplant.

- 9th February : The Board of Supervisors issues a use permit to PG & E; the use permit stipulates that the proposed facility could not be detrimental to the public health, safety, etc.; it also suggests that more specific details would be detrimental to the utility's plans.

- 24th May : The road is tentatively approved.

1961   - June : AEC reduces the cost of uranium by 34 per cent.

- 29th June : PG & E officially announces that the proposed powerplant will be nuclear.

- October : PG & E files with the California Public Utilities Commission (CPUC) for a "certificate of public convenience and necessity" (CPCN).

- December : The Board of Supervisors grants permits for the road.

1962   - 11th February : Harold Gilliam's article appears in the San Francisco Chronicle.

- 15th February : Army Corps of Engineers (Corps) holds hearings on the road, the first open hearings on any issue concerning the proposed powerplant.

- 7th, 8th and 9th March : The CPUC holds public hearings.

- March : Karl Kortum's letter to the editor appears in the San Francisco Chronicle.

- 30th March : The Corps approves the road.

- 21st and 23rd May to 6th, 7th and 8th June : Because of increased public attention, the CPUC is forced to reopen hearings on the issue of the powerplant certificate.

- June and July : Nuclear tests begin in Nevada.

- 17th August : The State of Utah orders milk destroyed and diverted from consumption by people because of possible contamination by iodine-131 from the nuclear testing in Nevada.

- 9th November : The CPUC grants an interim certificate to PG & E contingent on PG & E's allowing the public free access to as much of Bodega Head as possible and maintaining and correcting the levels of radiation in the surrounding water and air; the certificate is subject to review.

1963   - January : Excavation begins at Bodega and the road is built.

- CAPBHH sues.

- May : ACRS gives temporary approval.

- CAPBHH releases its report.

- 31st May : Lou Watters releases 1000 helium-filled balloons from Bodega Head to demonstrate possible drift of nuclear waste.

- 4th October : Scientists from the United States Geological Survey study the site at the request of the AEC, and find a new fault directly under the site chosen for the reactor.

1964   - 27th March : The Alaska earthquake occurs.

- October : Two committees of the AEC give conflicting reports.

- 26th October : The AEC rules that the plant will not be licensed.

- November : Construction stops, the cost having reached $4 million.

1972   - PG & E donates its property at Bodega Head for a state park.

b) Narrative

   While the Pacific Gas and Electric Company (PG & E, the "utility") did not publicly announce the Bodega Head powerplant until 1958, it had spent two years sounding out public officials in the area on their intentions, and a tacit pro nuclear campaign, as well as the buying up of land had begun as early as 1956. Officialdom in Sonoma County was highly responsive : the power plant would bring increased revenues into the county as well as substantially increase the tax base - a more tempting prospect than a state park, which would do neither. Prematurely optimistic, the county's Board of Supervisors began to consider the powerplant

37

a foregone conclusion, as did local businessmen and the local press, for Sonoma County, although situated near the populous San Francisco Bay Area, had been slow to develop and the business men were eager to see the county expand.

However, the attitudes of people in the town of Bodega Bay toward the plant were ambivalent. At a panel discussion organised by PG & E, some of the town's residents complained about the company's plans : landowners were outraged at the prospect of losing their land, and most did not want their remote area to be developed. But, poorly organised and unable to speak with a common voice, the people of Bodega Bay did not gain the attention of the rest of the county.

In the year that followed, appurtenances to the plant were being discussed and approved by county boards. The Board of Zoning Adjustments approved powerlines through Doran Park, the county's only park. This decision was the first tangible issue the opponents could rally around. Before the Board of Supervisors' approval, opponents had filed a petition, with over 1,300 names on it, demanding a hearing on the subject, but the Board paid little heed to the dissent and approved the powerlines, without a public hearing.

At the same time, the Harbour Commission prepared and submitted a plan for development of the harbour. It called for a new road to the plant. But instead of crossing over the headland, the proposed road would run along the edge of the bay, requiring the filling in of the tidelands. (It was later shown that this design had been promoted by PG & E. Trying to cut costs, PG & E proposed that the tidal basin be filled with the rock removed during excavation, to obviate hauling it away.) The proposal firmly cemented local opposition to the plant.

In the tidelands, there was a unique peat bog, the site of a prolific maritime ecosystem, and local people, especially the fishermen, were extremely concerned about its annihilation. On 24th May, 1960, after some minor and inconsequential adjustments, the Board of Supervisors approved the road but without a public hearing.

It would be hard to gauge the extent of the opposition. According to newspaper accounts of the time and to the position staunchly maintained by the Board of Supervisors, the opponents of the project were only a small "unprogressive minority" of the town's population. Since it had repeatedly been denied a public hearing, the opposition never became publicly manifest.

By January 1960, the request for a use permit for the plant was before the Board of Supervisors. Again, a public hearing was considered unnecessary. However, a final vote was delayed because "so many statements have been made in letters and petitions... that we (the Board of Supervisors) want time to check every possible segment of information available"(Santa Rosa Press Democrat, 22nd January, 1960). But county officials, eager to see the project completed, did little in the interim to evaluate opposition by the public. Identical resolutions, passed by two of the county's lower boards, recommended to the Board of Supervisors that it grant a use permit to the utility company. A mere five days later, in February 1960, the Board granted the use permit.

In what was presented to the public as a fait accompli, the Sonoma County officials had approved a powerplant whose design had never been publicly specified. Later, it would be revealed that they had approved the nation's largest nuclear reactor to date. The Board's haste was challenged by a State senator, James Rattigan.   At issue was the legality of approving the plans, or, better said, the "non-plans", of a plant of unspecified design. The Board had given what amounted to a blanket license to the utility. One of the conditions attached to the use permit was especially questioned by Mr. Rattigan. It read : "... the use for which application is made will not, under the circumstances of this particular case, be detrimental to the health, safety, peace, morale, comfort, and general welfare of persons residing or working in the neighbourhood" (Andersen, 1972). Even the county's counsel admitted that this condition would not be found legal in a court of law; however, he added, "We support your decision to act" (Santa Rosa Press Democrat, 10th February, 1960).

Designed to make decisions in the public interest, the county supervisors had acted with little apparent concern for public protection by confirming plans for a plant they had not yet officially inspected. This was tantamount to abrogating their role in the governing process. At the same time, they provided no me mechanism by which the public could respond. Without definite plans, no alternatives could be raised. Coupled to the scarcity of public knowledge was the public's lack of organisation : local citizens did not unite, and individually they were ill-prepared to fight the supervisors, whose propensities had been easily displayed.

After PG & E's plans were approved, the opponents of the plant were aided by an outside environmental group, the Redwood Chapter of the Sierra Club. The Redwood Chapter reviewed for the supervisors its desire to preserve Bodega Head since 1957.   They requested the Board to reconsider its decision. The supervisor's response was that the chapter's comments had come "a little late" in the process (Santa Rosa Press Democrat, 10th February, 1960).

Clarification of the issues was soon forthcoming. In 1959, the president of PG& E had announced that "an atomic powerplant will be built in one of the nine Bay Area counties" (Santa Rosa Press Democrat, 4th April, 1959); rumours began to circulate that the Bodega Head project was its intended site. Ben Guidotti, one of the Sonoma County supervisors, had come very close to confirming that himself in 1960, when he announced to the Chamber of Commerce in Bodega Bay that "the Pacific Gas and Electric Company is prepared to move ahead with its plans and start excavating for a plant site and intake pipe" (Santa Rosa Press Democrat, 22nd May, 1960). But the official announcement was not made until 20th July, 1961.

At that time, PG & E announced that it would build a General Electric prototype boiling water nuclear reactor in a "park" at Bodega. The "park" would eventually consist of two reactors capable of producing 325 MG of power. The original cost estimates were $61 million. Following the announcement on design, a vigorous discussion ensued within the nuclear community. The utility, with the backing of the nuclear-energy "community", was inclined to assert that Bodega would break the "economic barrier" and finally produce electricity from a nuclear reactor at competitive prices. Much of this was attributed to the superb location of Bodega Head.

In October 1961, PG & E filed with the California Public Utilities Commission for a certificate of "public convenience and necessity" facility. Without substantial alternative or conflicting testimony, the hearings were a "walk-away" for the utility.

But events soon occurred which turned the tide and aroused the public. Two articles, appearing within a month of each other in one of the area's largest papers, The San Francisco Chronicle, succeeded in expanding the public's interest. The first was Harold Gilliam's article in the "World" section of the paper. Far from being an entreaty, it eulogised the tragic loss of such a scenic spot. The second article was a letter to the editor by Karl Kortum. The letter was a political characterisation of the peremptory method used in acquiring, or rather usurping, the Bodega Head site.

The articles separated the case into two distinct issues. Gilliam's helped the public adduce exactly what and where Bodega was, raising the question, for the first time for many, of whether it was a suitable site. As the Sierra Club would later argue :

"The proposed reactor will take a priceless scenic resource for its site, but will produce nothing but common kilowatts" (SF /21,483, page 11). The Sierra Club,
expanding on this argument, demanded that the cost of
the plant include evaluation of the loss of the scenic
resource. They said, "The burden being on the applicant, they must show that its proposed use of a plant
site is, on balance, more beneficial to the public than
possible alternative uses. This is particularly true
where a plant site has been designated as a future
state park" (ibid., page 8). The Sierra Club was assuming the argument the Division of Parks ought to have
pursued.

Simultaneously, Karl Kortum's letter to the editor
of The San Francisco Chronicle clarified what had been
an abstruse understanding by the public of the unjust
proceedings in Sonoma. Angered by the more plutocratic
than democratic method of the proceedings, David
Pesonen, a member of the Sierra Club and a law student
at the time, created the California Association to
Preserve Bodega Head and Harbour (CAPBHH). The Sierra
Club had prevented Pesonen from becoming too much engaged in the fight; with the new group no such res-
traint on tactics remained. CAPBHH and the attention
drawn from the articles finally gave direction to the
scattered objections of the past.

CAPBHH, under the leadership of Pesonen, was a
professional group. It reached a membership of over
2,000 by the time the Bodega Head Case was over. Its
members were kept informed by a newsletter, and funding was provided by numerous "benefit" concerts and
private donations. By creating a mechanism and gaining
legal representation through a prominent San Francisco
law firm, CAPBHH was successful in continuing to direct
public attention to the case.

With the flood of renewed public interest in the
case caused by the two newspaper articles, the CPUC
was forced to reopen the hearings in May. In seven days
of public testimony and cross-examination, conflicting
tales were told. Attention had been directed to the
potential dangers of thermal and radioactive pollution
from the reactor. The plant was designed to cycle
through 950,000 litres of seawater a minute, returning
it to the ocean 9.4 degrees Celsius warmer than before.
In testimony scientists disagreed on the potential
effects this could have. A. Starker Leopold, speaking
for UCB, said that the university's scientists had
concluded that the plant's discharge "would not render
the (area) unusable for biological stations... (It would)
not even materially alter the site for biological studies" (Andersen, 1972). When the UCB group was asked
to be specific, it could only say that the effects
were uncertain at this point.

41

However, Joel W. Hedgpeth, a well known marine biologist from the University of the Pacific, was more willing to be specific. His claim, reiterated by other scientists, was that the large increase in temperature would substantially alter the marine environment. However opposite this testimony was, it had little impact in the face of the conflicting testimony of UCB, whose scientists ostensibly had much more at stake, since their marine biological laboratories were to be situated there. With their scientists noncommittal, the influence of other testimony was drastically reduced.

Two other issues received attention during the hearings. CAPBHH and the Sierra Club, along with individual citizens, challenged the right of the utility to take such a scenic site. Coupled with this argument was the danger of earthquakes. PG & E defended the suitability of the site for a nuclear reactor: its low and level ground, juxtaposed so perfectly to the cool waters of the ocean, and its remoteness made it a perfect site for a nuclear reactor.

Furthermore, PG & E offered convincing testimony on the importance of the nuclear programme in general, regardless of whether the reactor would break the "cost barrier". In regard to the risk of earthquakes, the PG & E engineers stated that they had employed the best seismologists to study the safety of the area. When pressed to produce the results of the seismologists' survey, PG & E refused. Finally, upon demand, PG & E complied, but after the public hearings, by filing the controversial "Exhibit 48".

Because critical cross-examination of other issues had resulted in a stalemate, debate over the earthquake hazard was crucial for the opponents. Unfortunately however, parties to the proceedings - unskilled in CPUC practice, and indeed, in legal procedure - were slow to insist upon their right to examine the document; also, they were slow to demand that PG & E's experts give their respective opinions under oath (Dissenting Opinion, decision Number 654701, 1963).

Considerable attention has been given to UCB's and the Division of Parks' unwillingness to enter the battle and defend their interests. With both groups keeping a "low profile" throughout the proceedings, it was difficult to offer persuasive counter testimony. It cannot be said that the utility or any one else put pressure on these groups to keep quiet. The Division of Parks did acknowledge - hesitantly - that the presence of the power-plant might inhibit subdivision and further development.

Their attitude cannot really be found at fault. On the other hand, UCB later admitted that scientists within the university had considerable doubts about the power-

plant and were adamantly against it. These reports never left the chancellor's office. UCG did not want to
go counter to PG & E's or the AEC's plans. At the time
(though this does not substantiate or disprove the hy-
pothesis), it is a fact that 56 per cent of the University's budget came indirectly from the AEC, in the
form of research grants. In November 1962, the CPUC
announced its initial decision and granted an interim
certificate to PG & E. Construction began shortly
thereafter at Bodega. The first rounds of the Bodega
Case had been lost by the environmentalists.

But the opposition had finally found its voice
in the dynamic leadership of David Pesonen. Had it not
been for the concerted efforts of his group, CAPBHH,
and for legal aid from the Sierra Club, the issue
might have remained a loss for the environmentalists.

But for Pesonen, the fight had only begun : on
the day after the decision by the CPUC, CAPBHH held a
meeting at Bodega. Detailed accounts of the "pros" and
"cons" of nuclear energy were given in an attempt to
countervail the pro-nuclear material that PG & E had
been distributing for years. The earthquake hazard
became the central concern of CAPBHH after it lost to
the CPUC. Originally, this issue had not been as important as the other previously mentioned issues.
Through the California courts, CAPBHH sought, in suit
after suit, to reverse the decisions of the state and
the county on legal grounds (see Annex II-1). Although
most of the suits were lost on higher appeal, the
needed public attention was gained.

In 1963, the Advisory Committee on Reactor Safeguards (ACRS), a committee within the AEC, conditional-
ly approved the proposed powerplant. With the original
proposal of a nuclear plant, the ACRS initially eval-
uates the quality and safety of the proposed reactor.
The ACRS wrote, "Tentative exploration indicates that
the reactor and turbine buildings will not be located
on an active fault zone. If this point is established,
the design criteria for the plant are adequate"
(Barbour, 1973).

"Dovetailing" the fault had always been a concern
of PG & E. In November 1959, the site of the plant had
been moved farther out onto the Head, ostensibly to
get away from the fault. PG & E had, in the meantime,
assured the AEC that, although the plant would be in
the general vicinity of the fault, it was well beyond
1.5 kilometres of the fault itself, and 300 metres
from the western edge of the fault zone.

Regulations on constructing a nuclear facility
on an active geological fault had been specified by
the AEC in a ruling in 1961. The ruling stipulated
that "no facility shoud be located closer than half

to a quarter mile (0.8 to 0.4 kilometre) from the surface of a known active fault". This criterion was later
liberalised in 1962 to simply 0.25 mile (0.4 kilometre)
in order (according to the then-director of the AEC)
to "eliminate the ambiguity of the earlier ruling"
(Pesonen, 1963).

The regulations were of special concern to PG & E
and to California in general. The San Andreas Fault,
the most active earthquake zone in the United States,
runs parallel to the coast of California for the entire length of the state, passing quite near Bodega
Head. The fault gained worldwide notoriety in 1906
as the cause of the great San Francisco earthquake
which left the city in a shambles. The fault was also
of special concern to CAPBHH.

The ACRS report had barely been announced when
another report emerged. The CAPBHH issued their report
that mada four contentions : (1) The reactor was within
0.25 miles (0.4 kilometre) of the San Andreas Fault,
not a mile (1.6 kilometres) away, as PG & E had claimed *; (2) CAPBHH experts had found evidence of an active fault underneath the reactor site **; (3) The
base of the reactor was not granite bedsock, as PG &
E claimed, but was a mixture of diorite, clay, and
sand, making the site unstable; (4) The summary report
given to the AEC by PG & E had suppressed and altered
initial reports and testimony of the PG & E staff,
including a vastly altered version of the "Exhibit 48"
being held by the CPUC (Barbour, 1973; David Pesonen,
personal communication).

In the face of these accusations and the urging
of Morris K. Udall, United States Secretary of the
Interior, the AEC was compelled to assess the problem.
Two independent United States Geologic Survey (USGS)
geologists went to Bodega. Their findings provided
sufficient evidence that a fault zone did indeed exist
near, or directly beneath the site selected for the
reactor. One of the geologists concluded : "Acceptance
of the Bodega Head as a safe reactor site will establish
a precedent that will make it exceedingly difficult to
reject any proposed future site on the grounds of extreme earthquake risk" (San Francisco News Call Bulletin, 4th October, 1963).

PG & E persisted in pursuing the site, however,
but the public was just as persistent. On 31st May,

---

* At that time, the exact location of the fault was open
to much conjecture among geologists.
** This was a misquotation of a Scripps Institution of
Oceanography report that had reported a fault some 30 metres
seaward of the plant.

1963, jazz trumpeter Lou Watters, an announced opponent of the plant, released 1,000 helium filled balloons from the peninsula. Attached to each was a tag that read, "This balloon could represent a radioactive molecule of strontium 90 or iodine-131 - tell your local newspaper where you found it." Some of the balloons drifted as far as populous San Rafael and Richmond, some 50 and 80 kilometres away, respectively.

Although PG & E was quick to denounce the results of Watters' demonstration as unscientific, public concern about the project was gaining momentum. CAPBHH published several articles describing the dangers of earthquakes and of nuclear energy. In what became know as the "Pure Milk Crusade", women picketed the local PG & E office. The possibility that radioactive particles could be randomly released (i.e. not as a consequence of an earthquake) also gained news.

Nature herself contributed to the compaign with a vicarious, but no less prognostic, event : in March 1964, the great Alaska earthquake occurred. The public's attention was not long in shifting from Alaska to the ground that most Californians were walking on. Public opinion changed its course almost overnight.

On 26th October, 1964, after having been silent for nearly a year, the AEC issued two conflicting reports. The ACRS again tentatively approved the plant, but the licensing division officially reported (Holdren and Herrera, 1971) :

> "In our view the proposal to rely on unproven and perhaps unprovable design measures to cope with forces as great as would be several feet of ground movement under a large reactor building in a severe earthquake raises substantial questions. It is our conclusion that Bodega Head is not a suitable location for the proposed nuclear power-plant at the present state of our knowledge."

Later, the AEC refused the licence.

In response to the AEC's refusal to issue a licence Robert H. Gerdes, president of PG & E, said (Barbour, 1973) : "We have repeatedly stated that if any reasonable doubt exists about the safety of the proposed Bodega Plant we would not consider going forward with it."

c) Analysis and Conclusions

In the Bodega Head Case, the regulatory agencies had difficulty in determining the scope of their responsibilities. Their refusal to consider certain issues (which they admitted were relevant to the case but beyond the scope of mandates) meant that those problems were either ignored or else deferred to

45

higher authority. For example, the Corps stated that passing on questions of aesthetics or marine ecology was beyond its responsibility, with the result that the true social costs of the access road were never fully assessed.

CPUC had the same problem. It never resolved the safety issues over the earthquake risk. CPUC allowed PG & E to submit a late document "Exhibit 48", which was never subjected to public scrutiny or the challenges of cross-examination. This was, according to Justice Bennett's dissenting opinion on the case, a fatal deficiency in the requirements of their service as a regulatory commission.

The safety issue was not resolved, nor was it completely assessed by CPUC; it was deferred to the AEC. CPUC's obligation to make a complete study was never fulfilled, much as Sonoma County's Board of Supervisors had never required a complete detail of the plant.

A complete study of PG & E seismic survey, "Exhibit 48", would have shown that the seismologists had disagreed among themselves about the suitability of the site. Had CPUC acted in accordance with its mandate, and for the people of California, it would have had to cancel the project because of the uncertain risk factors (Bennett, 1963). The burden of proof was on the utility, which was required to demonstrate both the need and the feasibility of the project. In the Bodega Head Case, CPUC accepted PG & E's testimony at face value over others' testimony, without demanding fuller or more detailed examination.

There are several reasons for this. The CPUC depended on PG & E to provide most of the technical information. Beyond that, it depended upon the chance abilities or inabilities of concerned parties to provide more extensive or conflicting reports. There is no way CAPBHH or the Sierra Club could have prepared as extensive a report as the PG & E staff was capable of preparing.

PG & E gave assurances that the best scientists had been chosen to determine the hazards. Where there was divergence of opinion, CPUC chose either not to become involved in the problem or opted for the carefully prepared testimony of PG & E. Several issues were contended at the CPUC hearings. The possible effect of thermal and radioactive pollution on the marine biota was raised. PG & E, with the tacit support of UCB, asserted that the effects would be minimal or undeterminable. Against the argument, the environmentalists' testimony, which was the anti-thesis of the utilities', carried little weight.

Another of the CPUC's functions was to compare

the social costs of a proposal with its benefits.
Again, the CPUC had few criteria by which to judge.
The utility outlined succinctly the anticipated growth
and demand for power. On top of this, they added the
benefits of growth and construction in Sonoma County;
these were tangible social benefits. The Sierra Club
challenged PG & E benefit-cost analysis by asking PG
& E to incorporate the costs of relocating the facility, but this was never required by the CPUC.

Against the benefits stood the less tangible costs.
How could CPUC really evaluate the costs of losing a
scenic resource ? Yet the environmental groups asked
that this be weighed against the proposed benefits of
the plant. But finding a reliable basis for comparison
was difficult, if not impossible. The public, well
represented on this point, could not argue in specific
terms nor quantify their position, while PG & E could.
The CPUC never required PG & E to more fully examine
alternative sites.

The failure of the environmental groups to per-
suade the CPUC was primarily due to their lack of legal and organisation expertise. Their concerns were
never properly articulated, nor were their legal opportunities fully recognised. However, after the CPUC
hearings and the subsequent decision, David Pesonen
and the CAPBHH persisted through the channels open to
them. Although they got involved in the process later,
their efforts raised public attention and interest in
the case.

Continued lawsuits and other court action, although
often successful, helped to keep the Bodega Head Case
in the news. This had the same effect as the two
articles had had before. By drawing attention to certain issues, the environmentalists kept them alive;
the issues therefore were not overlooked. The attention given to the danger of possible earthquakes, the
promulgation of pamphlets, the holding of seminars,
and the creation of newsworthy events, heightened
the public's demand that the government be certain
about the safety of the plant. The discrepancies in
the documents and the presence of geological faults
might have been overlooked by the entire AEC had it
not been for the efforts of the local environmental
groups involved. After all, the ACRS had recently
approved the plants on the assurance of PG & E.

4.   The Kaiparowits Plateau Coal-Fired Electricity
     Generating Plant

     a) Chronology

     A chronology of the major events in the Kaiparo-
wits Case follows.

1950s	- WEST plans a 36,000 MW power complex near the "Four Corners".
1963	- The Four Corners plant begins operations.
1964	- Three power companies outline the Kaiparowits proposal to the Utah Water and Power Board.
1966	- Waterrights are sought for cooling the plant; a complexe interstate water allocation system results in three years of negotiations.
1969	- Water rights are granted by Secretary of the Interior Hickel. - NEPA passes, requiring environmental impact statements (EIS's).
1970	- The Sierra Club sues to bar the federal government from co-operating in the development of the powerplants until the department of the Interior (DOI) agrees to produce an EIS, as required by NEPA. - The Clean Air Act passes, giving strict protection to clean-air regions.
1971	Secretary of the Interior Morton announces a year-long moratorium on the approval of power-plants; the DOI begins its Southwest Energy study.
1973	- Applications for rights-of-way are submitted; they are rejected by Secretary Morton on environmental grounds; because of changes in siting and design, he agrees to reconsider. - A new EIS is required, to be prepared by agencies of the federal government pursuant to a ruling of the United States District Court of Appeals in 1972.
1975	- The draft EIS is completed in February; public hearings delay its acceptance. - November : The Sierra Club files a petition with the California Public Utilities Commission, requesting it to determine whether the plant is needed. - December : The utilities announce a delay, citing "objections by environmental groups and lengthy approval processes".
1976	- March : The final EIS is issued. - 14th April : The utilities withdraw their applications because of "economic infeasibility".

b)   Introduction

In the mid-1950s, the major cities of the south-
western United States were expanding rapidly. In response, a group of power companies ("utilities") in the
region, projecting a 6.5 per cent annual increase in
demand for electricity (Los Angeles Times, 15th April,
1976, page 1), began to seek a good site for a system
of powerplants. Their search ended at the so-called
"Four Corners" area, where the states of Utah, New
Mexico, Arizona, and Colorado meet. There, a thick,
untapped seam of coal lay near the surface, and the
utilities expected minimal resistance to the presence
of a system of coal plants because the region was spar-
sely populated, mostly by poor Navajo and Hopi Indians.
Water for cooling the plant would be scarce, but could
be brought from the Colorado River and its tributaries
(Holdren and Herrera, 1971).

The Four Corners region also contains some of the
most spectacular scenery in the United States, and
conservationists and other Americans in love with the
land had struggled to preserve it for the enjoyment of
future generations.

The scenic value of the area had caused much of
it to be designated as national parks, national monuments, and other protected sites, such as Bryce Canyon,
Zion, Canyonlands, the Grand Canyon National Parks,
and Capital Reff, Arches, Cedar Breaks, Natural
Bridges, Canyon de Chelly, and Rainbow Bridge National
Monuments (ibid.). One-fifth of the land administered
by the National Park Service (NPS) is situated within
400 kilometres of the Four Corners region.

Twenty-three utility companies formed a consortium
called "Western Energy Supply and Transmission" (WEST).
WEST developed plans for what would be the United
States' largest powerplant complex, capable of producing
36,000 megawatts (MW) of electricity in the 1980s. The
six largest powerplants were to be : Four Corners
(2,075 MW) and San Juan (990 MW), on the San Juan River
near Farmington, New Mexico; Navajo (2,310 MW) and
Kaiparowits (5,000 MW), on opposite sides of Lake
Powell - one in Arizona and the other in Utah; Mohave
(2,310 MW), on the Colorado River in Nevada; and Huntington Canyon (2,000 MW), on Huntington Creek in
central Utah (Holdren and Herrera, 1971, pages 161
and 162).

Very few environmentalists knew about the powerplant complex and its potential effects until the Four
Corners plant began operating in 1963. The plant was
one of the worst polluters in the nation, spewing 300
tonnes of particulate matter into the air each day -
more than is allowed in New York City and Los Angeles
combined. An investigation showed that the operator

49

of the plant had failed to install two-thirds of the pollution-abatement equipment specified in its contract. New Mexico ordered the plant to clean up or shut down by the end of 1971 (ibid., page 163).

Since it was too late to halt the powerplants already in operation, conservationists concentrated their efforts on Kaiparowits, the largest plant proposed. Efforts by such local groups as the Central Clearing House and Black Mesa Defense Fund publicised the environmental effects of the whole complex of plants, shocking the public. Life magazine published the article, "Hello Energy - Goodbye Big Sky". However, there was no major environmental opposition until 1970 (ibid., page 166).

The Kaiparowits project was destined to repeated delays. The utilities sought water rights to Lake Powell for cooling water in 1966. But the lake's source is the Colorado River, which is governed by a complex inter-state water allocation system (The Wall Street Journal, 7th September, 1976, page 1). Three years of negotiations ensued over the Department of the Interior's (DOI's) desire to guarantee that there would be enough water from Lake Powell for states in the lower (southern) Colorado River Basin in case of drought (Deseret News, 15th April, 1976). Finally, on 3rd October, 1969, Secretary of the Interior Walter J. Hickel signed documents granting water rights (ibid.)

At about this time, the environment was becoming a major issue nationally. The National Environmental Policy Act of 1969 (NEPA), which required the preparation of environmental impact statements (EIS's) was passed by the Congress in December 1969 and signed by the President on 1st January, 1970. The Clean Air Act of 1970 ensured strict protection for already pure air (ibid.). There was increasing recognition that local developments could have extensive regional consequences. The events at Kaiparowits would now be receiving the attention of a national audience.

In 1970, the Sierra Club sought to enjoin the federal government from co-operating with powerplant developers, charging that the agencies involved had failed to issue EIS's on the effects of various water-withdrawal permits and grants of right-of-way, as required by the newly passed NEPA. The Club's suit was dismissed in 1972, but had served to stall development of the plant (Uinta News, May 1976, page 1; Holdren and Herrera, 1971, page 165).

Environmental concerns were raised in other quarters as well. For example, in May 1971 the Senate Interior Committee held a week of public hearings in the affected states to see whether it could help resolve the problems.

It could not, but the dilemmas were defined (Holdren and Herrera, 1971, pages 166 and 167) :

"Executives of the electric utilities presented their case by emphasizing the benefits of having great blocks of cheap power available. They argued that electricity creates employment, thus causes a higher standard of living... More electricity is needed, among other things, to save the environment by powering sewage treatment plants, mass transit, junk compressors, recycling processes... Moreover, WEST's new plants would be outfitted with the very latest in pollution abatement devices and so would meet all applicable air quality standards.

But the plants are so big, countered many environmentalists, that even if they had the best safeguards, they would still emit 240 tonnes of fly ash every day, plus 2,160 tonnes of sulphur dioxide and 1,350 tonnes of nitrogen oxides... By withdrawing more than 80 billion gallons annually from the Colorado River Basin, the plants would also increase the salinity content of the water flowing downstream to thirsty southwestern farms. And then there were the virtually permanent ravages of strip mining to consider."

During the hearings, a coalition of conservation groups took full-page advertisements in The New York Times, the Los Angeles Times, and other national newspapers, to protest the devastation of the Southwest. Attention to the advertisements was guaranteed by their headline : "Like Ripping Apart St. Peters, in Order to Sell the Marble" (ibid., page 166). On 27th May, 1971, Secretary of the Interior Rogers C.B. Morton announced a year-long moratorium on the approval of powerplants. During the next two years, DOI, through its Southwest Energy Study, investigated Kaiparowits and similar projects (Deseret News, 15th April, 1976).

Trying to reduce opposition by environmentalists, the power companies decided to propose building the plant at a new location, about 19 kilometres farther from the Navajo Plant. In addition, they reduced the size from 5,000 to 3,000 MW, and proposed that improved pollution controls be added (Deseret News, 15th April, 1976, and The Wall Street Journal, 7th September, 1976, page 1). Applications were submitted in 1973, but were rejected by DOI on the basis of its Southwest Energy Study. However, Governor Rampton and other prominent people from Utah convinced DOI to reconsider its decision, in light of the changes in siting and design (Deseret News, 15th April, 1976).

But in the meantime, a 1972 federal court of appeals had ruled that EIS's must be prepared by

51

agencies of the federal government. Thus the Kaiparowits report had to be prepared by ten federal agencies, resulting in another two-year delay (The Wall Street Journal, 7th September, 1976, page 13.)

In February 1975, the draft EIS was issued and circulated for a year among concerned parties and federal agencies (The Wall Street Journal, 7th September, 1976, page 13). The Environmental Protection Agency (EPA) virtually rejected the EIS, claiming that it was too narrow in scope and urged DOI to postpone any decision (Uinta News, May 1976, page 1).

Public hearings served to delay its acceptance, and revealed that "the effects of the already operating Navajo Plant were not yet known" (Audubon, March 1976, page 74). The EIS reported that "there is enough evidence from the one unit operating at the Four Corners plants to cause grave doubts on the possibility of clearing up the emissions to a point where they will not have a serious adverse visual impact" (Sierra Club Bulletin, August and September 1976, page 26).

At this time, "grassroots" efforts of conservationists intensified. Numerous leaflets were distributed and mailed to the public (Ron Rudolph, FOE, personal interview). Workshops and conventions were sponsored by coalitions of conservation groups. At one conference, Kaiparowits was declared "a waste of money, a waste of energy, and a waste of significant scenic resources" (Deseret News, 10th January, 1976, pages A3 and A8). These events were publicised and reached a national audience. One delay, ostensibly made to permit further study or rerouting the power-lines, was attributed by Utah's junior senator, Jake Garn, to DOI being "so damn afraid they are going to be sued" (Sierra Club Bulletin, August and September 1976, page 7) by the environmentalists.

In November 1975, the Sierra Club filed a petition with the California Public Utilities Commission (PUC), asking that the PUC take jurisdiction over the Kaiparowits powerplant, and review independently the alleged need for its construction, stating (Uinta News, February, 1976, page 1) :

"Until now, no Government agency has reviewed whether this expensive and environmentally destructive project is necessary in light of California's future energy demands, and the utilities have escaped regulation merely by locating the facility out of state."

The petition requested that the PUC require the two California utilities to obtain a "certificate of public convenience and necessity" before proceeding with the project. This action resulted in an announcement in March 1976 that California regulatory officials have concluded that hearings should be held to determine whether the power from Kaiparowits would be needed in California (Rocky Mountain News, 17th March, 1976, page 40).

On 30th December, 1975, the utilities announced another delay, blaming "objections by environmental groups and lengthy approval processes" (Deseret News, 17th April, 1976). It was realised that, even if DOI approved the project, "they would still have to obtain more than 220 permits and authorisations in a grinding journey through 32 federal, state and local agencies" (The Wall Street Journal, 7th September, 1976, page 13).

By the time the final EIS was issued in March 1976, the NPS had conducted studies that demonstrated that odours produced by the plant would be detectable over 160 kilometres away (The Wall Street Journal, 7th September, 1976, page 13). The NPS also announced that it would conduct a study of air quality in three neighbouring national parks (Audubon, March 1976, page 90). Classification of the air in the parks as "Class I" would probably prevent the construction of coal-conversion plants in the area.

By April, meeting the requirements of NEPA had delayed construction for six years. The estimated cost of building a plant half as big as that planned in 1965 for $500 million was $3.7 billion. The costs continued to rise at the rate of over $1 million per day, and the plant's acceptance was still in doubt (The Wall Street Journal, 7th September, 1976, page 13).

On 14th April, 1976, the utilities withdrew their applications because of a series of uncertainties, particularly those relating to ultimate costs (Los Angeles Times, 15th April, 1976, page 1). Among the uncertainties were :

1. An amendment pending in Congress that would tighten air-quality standards near national monuments;

2. Greater interest by the CPUC and the California Energy Commission to supervise the out-of-state construction of electric powerplants; and

3. A suit recently filed against the project on environmental grounds and the possibility that more suits would come (Los Angeles Times, 15th April, 1976, page 1).

When announcing the decision to drop the proposed plant, William R. Gould, executive vice-president of

Southern California Edison, said the plan had been "beaten to death by the environmental interests" (Washington Post, 15th April, 1976).

c) Analysis

The goal of the environmental groups involved in the Kaiparawits Case was to halt construction of a poorly sited powerplant. The major factor that aroused environmental opposition was the location of a 5,000-MW, coal-fired electricity-generating plant within 400 kilometers of some of the most scenic areas in the United States. The environmentalists were concerned most about of air pollution that would be caused by such a gigantic plant. The draft EIS has stated (Sierra Club Bulletin, August and September 1976, page 26) :

"If this visual pollution constantly drifts into the nationally and internationally important scenic areas such as Grand Canyon, Rainbow Ridge, Lake Powell, Zion Canyon, Bryce, Arches, Canyonlands, etc., the effect of the panoramic viewing values could be catastrophic. There is enough evidence from the one unit operating at the Four Corners plants to cast grave doubts on the capability of clearing up the emissions to a point where they will not have a serious adverse visual impact."

Another consideration, which emerged in 1975, was the realisation that the power to be provided by the Kaiparowits powerplant might not be needed. That utilities involved in the Kaiparowits proposal did not have the proper perspective for objective forecasting was recognised by the Federal Energy Administration (FEA) when it commented (Audubon, March 1976, page 76) :

"Those projections which have been made either are insufficiently detailed or rest on assumptions considered too speculative as a basis for planning... Recognising that demand forecasts must necessarily reflect subjective judgements, FEA believes it important that there be forecasts compiled independently of those produced within the electric utility industry."

The truth of that statement was demonstrated in the present case (Audubon, March 1976, pages 76 and 84) :

"when it was revealed that a staff study made the previous November (1975) for California's Energy Resources Conservation and Development Commission but suppressed by the commission's development-minded majority, had reported, It may be possible to delay the Kaiparowits project without suffering any loss in electricity needed to satisfy demand in the southern California market areas served by the two participating utilities."

These factors, added to the growing environmental movement, resulted in lengthy disputes over acceptance of the powerplant proposal. Environmentalists sought to halt the plans. Some of the environmental groups involved and the tactics they employed to achieve their goal will now be examined in light of the controversy on the Kaiparowits proposal.

During most of the Kaiparowits Case, those who generally favoured development of the powerplant included the power ("the utilities") companies (Southern California Edison, San Diego Gas and Electric, and Arizona Public Service Company), the Federal Energy Office (FEO); and Governor Rampton of Utah and most citizens of Utah, who wanted the increased tax base and employment the project would bring.

Those who sought to halt construction of the facility included the Sierra Club, Friends of the Earth (FOE), The Wilderness Society (TWS), the Environmental Defence Fund (EDF), and National Wildlife Federation (NWF), the American Rivers Conservation Council, Arizonans for Quality Environment, the Colorado Open Space Council, the Council on Utah's Resources, The Desert Protective Council, the Ecology Center of Southern California, the Environmental Policy Center (EPC), the Federation of Western Outdoor Clubs, the Four Corners Wilderness Workshop, Interested in Saving Southern Utah's Environment, the National Parks and Conservation Association (NPCA), Trout Unlimited, the Alpine Club, NPS, EPA, and DOI (Denver Post, 31st August 1975, page 37).

The last three organisations are agencies of the federal government, not citizens' groups; therefore, it would be difficult to distinguish the influence of private citizens from that of federal agencies. Indeed, the conjunction of two sources of influences may have lead to results different from those of only one source.

Litigation played an important role at two points in the Kaiparowits Case. The suit brought by the Sierra Club in 1970 concerning EIS's was dismissed in 1972, but it was one of a series of delays that eventually halted the project. The Sierra Club's petitioning of the CPUC in 1975 increased the CPUC's interest in examining the need for the proposed development, which was one of the uncertainties the utilities cited when they announced their decision to withdraw their applications. They also cited their fear of further environmentalist suits.

The attempts of environmental groups to enact legislation were influential and almost decisive in the

Kaiparowits Case. With the Kaiparowits proposal directly in mind, they proposed amendments to the Clean Air Act of 1970   that almost passed, but that were finally killed by the Utah senators (Ron Rudolph, FOE, personal communication). The amendments would have required the mandatory designation of much of the parklands near Kaiparowits as Class I, ruling that any pollution would have "significant impact". Had the amendments passed, the Kaiparowits proposal would have been rejected automatically.

In 1975, after the draft EIS was issued, many members of environmental groups attended public hearings at which they demonstrated the magnitude of their concerns and fairly represented their interpretations of the issues. For example, the constant pressure they put on Russell E. Train, Administrator of EPA in 1975, persuaded him to oppose the immediate development of Kaiparowits (Ron Rudolph, FOE, personal communication). Also, public hearings on the draft EIS served to delay acceptance of the proposal and revealed that "the effects of the already operating Navajo plant were not yet known" (Audubon, March 1976, page 74), implying that the effects of Kaiparowits could not be predicted.

Further, constant efforts were made to keep all actions relating to Kaiparowits in the news. In 1971, "a coalition of conservation groups took full page ads in The New York Times, Los Angeles Times, and other newspapers to protest the devastation of the Southwest. The headline... 'Like Ripping Apart St. Peter's, in Order to Sell the Marble' guaranteed attention" (Holdren and Herrera, 1971, page 166). Environmental groups obtained lists, and with a constant flow of leaflets kept Kaiparowits in the public spotlight (Ron Rudolph, FOE, personal communication). The magnitude of the controversy made each hearing and other developments newsworthy events, so that the happenings in southern Utah were publicised in such national newspapers as the Los Angeles Times, The Wall Street Journal, the Denver Post, and the Washington Post. Coalitions of environmental groups held conferences and issued statements declaring Kaiparowits to be a "waste of money, a waste of energy and a waste of significant scenic resources" (Deseret News, 10th January, 1976, pages A3 and A8).

These public-awareness tactics kept the Kaiparowits controversy in the news and kept the public informed about the negative impacts that had to be compared with the social benefits of the proposed powerplant. This resulted in a more complete examination of the proposal.

d) Consequences of the Kaiparowits Controversy

Before the proposal was cancelled, the power companies had attempted to change the siting and design of the plant. In their 1973 application, they proposed that the plant be situated 19 kilometers from the original Nipple Bench site near Lake Powell. In addition, they proposed that the plant produce 3,000 MW of electricity instead of the 5,000 MW originally planned, and that additional pollution-abatement devices be added. These proposed alterations demonstrate the pressures the utilities were under, and were an effort to "reduce environmental opposition" (The Wall Street Journal, 7th September, 1976, page 1).

Although the proposed amendments to the Clean Air Act did not pass, they were studied very seriously. If they had passed, the intensive lobbying efforts of groups such as FOE and the Sierra Club would have had extensive ramifications. The mandatory designation of the nearby NPS areas as Class I would have greatly diminished the chances for development in regions of spectacular natural beauty.

On 17th March, 1976, CPUC officials concluded that "hearings should be held on whether two California utilities need electricity from the planned Kaiparowits powerplant in southern Utah. The recommendation came from the Sierra Club last November" (Rocky Mountain News, 17th March, 1976, page 40). Before this decision was made, no regulatory agency had expressed an interest in determining whether the proposed powerplant was needed. The decision set a precedent for future out-of-state developments from which the electricity produced would be sent to California.

The most evident effect of the events in the Kaiparowits Case was the eventual cancellation of the powerplant proposal. When the executive vice-president of Southern California Edison, William R. Gould, announced the decision to withdraw applications for the plant, he said that the plan had been "beaten to death by the environmental interests (Washington Post, 15th April, 1976). Those "environmental interests" included not only environmental groups, but also certain agencies of the federal government such as the NPS, EPA, and DOI, part of whose responsibility is to protect the nation's environment.

The NPS conducted and publicised studies of odours caused by the plants in nearby national parks and recreation areas. EPA had virtually rejected the 1975 draft EIS and urged DOI to postpone a decision. DOI had refused to approve the power companies' plans in 1973, on environmental grounds, and in 1975 had delayed the project further to allow additional study of the feasibility of rerouting the powerlines.

However, the uncertainties cited by the utilities in April 1976 focussed on amendments to the Clean Air Act that were pending in Congress, greater interest of the PUC in supervising construction of the plant, and the possibility that more suits would be brought by environmentalists. It is most likely that neither environmental groups nor federal agencies alone were responsible for preventing the development at Kaiparowits; rather, it was their combined actions that caused the successive delays which led to cancellation of the proposal.

Other consequences of the Kaiparowits Case may not become apparent for some time. The public's knowledge about the issues involved may lead to more-informed decisions on the development of power facilities in the future. It was observed in 1976 that "Growing numbers of Utah's citizens are sensing the catastrophe that Kaiparowits would mean to their state and are joining in opposition to it with aroused conservationists throughout the nation who regard the great parks and canyonlands as an American heritage that must be saved" (Audubon, March 1976, page 90).

Hearings such as those held by the Senate Interior Committee in May 1971 demonstrated the magnitude of the issues involved, and although the problems were not solved, the hearings helped to define the dilemmas. Perhaps these, together with future controversies that raise similar issues, will provide impetus for further legislation on standards and procedures for protecting the environment.

5.   The Midland, Michigan, Nuclear Powerplant

    a)   Chronology

A chronology of the major events in the Midland Case follows.

1967   - Consumer's Power Company (CPC) purchases 240 hectares of land near Midland, Michigan.

1969   - 13th January : CPC files for a construction permit with the AEC.

    - September : CPC re-evaluates its construction schedule and sets a new one.

1970   - May : CPC requests a Limited Work Approval (LWA) from the AEC, in order to begin clearing the site and making minor preparations for the foundation.

    - 18th June : ACRS completes its review of the plant, which it approves in a letter to the AEC; in the letter, ACRS states that some of

the problems discussed would have to be resolved by the Applicant and ACRS, but that the
plant was acceptable.

- July : Wendell Marshall inquires of the AEC
whether the cooling pond could cause fogs,
becoming thereby the first person on record
to question the proposed plant.

- 30th July : The AEC grants the LWC to CPC,
stipulating that this action will not influence
its later decisions.

- August : The Michigan Water Resources Commission hears testimony before it issues a
permit for the plant; Mary Sinclair stresses
the danger of the plant.

- October : AEC names an Atomic Safety and
Licensing Board (ASLB) for Midland plant; its
members are : Clark Goodman, David Hall, and
Arthur Murphy (chair).

- 27th October : ASLB hearings on the Midland
proposal are announced.

- November : Petitions to intervene are accepted.

- 17th November : A prehearing conference is
held.

1971   - Throughout the year, the hearings continue
in Midland.

- July : Calvert Cliffs' case is decided by
the Court of Appeals.

- September : CPC files an Environmental Impact
Report (EIR).

1972   - January : The ASLB circulates a draft EIS
to state, federal, and local agencies, and
interested parties, for their comments.

- March : The final EIS is completed.

- 17th May : Evidentiary hearings on the EIS
are scheduled.

- Spring : Hearings on the plant's emergency
core-cooling system (ECCS) begin in Washington,
D.C. (Bethesda, Maryland).

- 15th June : The ASLB's hearings on the EIS
end.

- 17th December : In an initial decision, the
ASLB recommends issuance of a construction
permit.

1973   - 7th January : Saginaw Intervenors move to recall and revoke the initial decision, alleging bias.

- 15th January : Saginaw files Exception to the Initial Decision with the Appeals Board.

- 1st February : The motion of 7th January is denied.

- 26th March : An Interim Judgement of Exception is given; "quality assurance" and "quality control" are ruled necessary amendments to the decision.

- 18th May : The Appeals Board rules on the remaining Exceptions filed by Saginaw and finds "no occasion to alter the results reached in the initial decision".

- May : Myron Cherry and Anthony Roisan file a suit against the AEC to halt all construction of reactors in the United States because they are unsafe; the case becomes famous as "Ray versus Nader".

- Another suit is filed, by Business and Professional People for Public Interest and Friends of the Earth (FOE), in a successful effort to force release of the second "Brookhaven Report".

- 10th July : The AEC refuses to review the initial decision.

- November : The AEC reverses its previous ruling and declares that energy conservation must now be taken into account (in re Niagara).

- December : Saginaw Intervenors requests clarification of the Midland plant in light of the Niagara decision.

1974   - January : The Court of Appeals orders the AEC to respond to the request filed by SAGINAW; before this order, the AEC had refused to present a clarification.

- 24th January : The AEC responds by stating that Niagara was not retroactive.

- February : CPC and Dow renegotiate their contract and significantly reduce Dow's dependence on the plant.

- February : Saginaw requests a reopening of the hearings because of the change in Dow's obligation in the new contract.

- 2nd February : Saginaw's request is denied.

- 11th April : After calling for all relevant contentions, the AEC again decides not to re-open the hearings.

- May : CPC asks the ASLB to defer indefinitely the hearings on another powerplant, on the grounds that sales of electricity were substantially lower than had been anticipated.

- 27th November : Saginaw Intervenors argues its case for a remand of the hearings before the Court of Appeals in the Washington, D.C. Circuit.

1975   - January : President Ford signs the Bill abolishing the AEC and creating the Energy Research and Development Agency (ERDA) and the Nuclear Regulatory Commission (NRC).

- The Court of Appeals puts the Saginaw Case in abeyance, pending resolution of related case ("NDRC versus NRC").

b)   Narrative

The Midland Case exemplifies many of the problems which occur at present during the arduous and lengthy process of licensing a nuclear reactor. A distinctly new posture emerges in the actions and method of the environmental groups since Bodega. With the Passage of the National Environmental Policy Act of 1969 (NEPA), concerned citizens and environmental groups are now capable of demanding a substantive review of the proceedings. At the height of the Midland Case, the courts ruled favourably, from an environmental standpoint, on the "Calvert Cliffs" Case. Later in the process, the AEC was split into two entities, the Energy Research and Development Agency (ERDA) and the Nuclear Regulatory Commission (NRC). As the case continues, new rulings draw out the process. New precedents, some set by the Midland Case itself, further embroil what has been one of the longest and most vehemently contested nuclear debates of all time.

Midland, a town of approximately 41,000 people, is situated in the largely rural, middle-eastern section of the state of Michigan, approximately 100 kilometers southwest of Lake Huron.

In 1967, Consumer's Power Company (CPC), one of Michigan's largest power companies ("utilities"), purchased 260 hectares of land near the town, on the banks of the Tittabawassee River. Two years later, CPC announced that the site had been chosen for a nuclear powerplant facility. The plans were for two pressurized water reactors capable of generating 1,300 megawatts (MW) of electricity at a cost of $400 million.

The site had been chosen for two reasons : (1) its proximity to the Dow Chemical Company's plant, and (2) the flatlands nearby.

The flatlands would be used to construct a 530-hectares, man-made cooling pond to dissipate the heat from the reactors into the atmosphere. Dow Chemical, directly across the river, agreed with CPC to purchase 400 million pounds of steam per hour to be used in the chemical processing. This would allow the chemical company to dismantle its bulky and outdated fossil-fueled boilers that supplied the steam. Dow also agreed to purchase large amounts of electricity from the proposed plant.

"Colocation", such as this, has been a goal of environmentalists for quite a while because it promises better use and allocation of resources. The coordination of location and the exchange of resources as proposed by Dow's and CPC's joint venture assured a more competent use of resources and was a tremendous improvement in the management of resources. As with all nuclear reactors, a large proportion of the heat generated to produce the steam that turns the turbines is waste; it must be disposed of. The proposed arrangement at Midland would have turned the waste heat into a valuable resource by using it in Dow's chemical-processing operations.

CPC was careful to prepare people in the town of Midland and the surrounding region for its proposed project, by offering seminars, guest speakers, and information describing the benefits of nuclear power. This pronuclear campaign was tremendously successful; local influential citizens, as well as numerous organisations, fell in firmly behind the proposed project. As in the Bodega Head Case, the prospect of increased tax revenues and of economic growth was decisive.

When Dow announced its interest in the plant, little additional persuasion was necessary. Midland is above all a company town. "It has been said often that Midland undergoes a severe shaking up whenever Dow Chemical Company sneezes" (Midland Daily News, 3rd December, 1976, page 1). Dow's firm commitment to the plant, plus its assurance of the safety and reliability of nuclear energy, convinced the local citizenry of its benefits. Furthermore, rumours, with a grain of truth to them, circulated that Dow would have to move the plant if the reactors were not built. This was partially true because Dow's old boilers were under scrutiny from the Michigan State Air Pollution Board and would soon have to be renovated, altered, or removed. The town, heavily dependent on Dow for its livelihood , also completely trusted Dow's scientific expertise : if Dow engineers said the plant was safe, most people in Midland believed it would be.

Attention shifted, however, in 1970. Mary Sinclair, a writer loosely affiliated with Dow, noticed serious ambiguities in the technical information she had seen on nuclear energy in general and on the Midland plant in specific. In testimony given before the Michigan Water Resources Commission on 10th August, 1970, she outlined several of the unresolved safety issues surrounding nuclear reactors. She also pointed out the "questionable attitude of the AEC toward public health and safety".

Impatient to begin construction, CPC filed for an exempting LWA to begin clearing the site of construction. In July 1970, the AEC granted the LWA, adhering to the guidelines of the Atomic Energy Act by stipulating that the decision would in no way affect its later decisions. By its own decision, CPC had committed itself to construction before it knew how long the licensing process could take. Although CPC had several past cases to judge from - cases that would have persuaded them to proceed differently - CPC felt confident that the procedure would be a quick one. CPC had underestimated the length of the licensing proceedings and - it might be said - had disregarded their importance as well. Later, faced with delays and the construction cost, some due to the protracted hearings, they realised their error. Clearly, however, they had committed themselves to a troublesome position.

Notice of the ALSB hearings was issued on 27th October, 1970. Before that, the AEC had named the members of the ASLB. Petitions to intervene were filed by several groups after the notice. The five major groups to take part in the proceedings were : Midland Nuclear Power Committee (MNPC), the Saginaw Intervenors ("Saginaw"), Mapleton Intervenors ("Mapleton"), the Environmental Defense Fund (EDF), and Dow Chemical Company ("Dow"). Together they represented a broad spectrum of goals and motives.

In favour of licensing the plant and seeing the licensing procedure proceed expeditiously, were MNPC and Dow. MNPC was a local group of fifty people representing "trades, professions, businessmen, organisations and industries". Their chairman was Dr. Wayne North of the First Methodist Church of Midland. They closed their petition to intervene with the remark, "... it is in the best interest of all citizens of the area surrounding the proposed plants that the board render a favourable decision" (Petition to Intervene, 10th November, 1970, page 5). Dow also intervened - as could be expected, in support of the plant.

Expressing a concern to see the safety of the reactor assured were two groups. EDF, a non-profit corporation made up of "scientists and other citizens

dedicated to the protection of man's environment",
filed to intervene, with the intention of guaranteeing
"full exploration of the impact of the project on the
environment and alternatives to it" (Petition to Intervene, 12th November 1970, page 3).

The second group was Saginaw, a conglomeration
of citizens' groups who were filing under a common
petition.

Saginaw consisted of seven groups : The Saginaw
Valley Nuclear Study Group; the Citizens Committee
for Environmental Protection of Michigan; the Sierra
Club; the United Auto Workers; Trout Unlimited; the
West Michigan Environmental Action Council; and the
University of Michigan Environmental Law Society.

The Saginaw Valley Nuclear Study Group intervened
through Saginaw because it was "interested in the
dissemination of information and stimulation of public
awareness and involvement on the study of nuclear
power". The expressed concern was that "the construction and operation of the Midland Plants will result
in an unsafe and inefficient exercise in the use of
atomic energy" (Petition to Intervene, 12th November,
1970, page 2); the Citizen's Committee for the Environmental Protection of Michigan intervened because its
members were concerned that "their lifestyles, their
occupations, and their investments, both emotional and
financial, will be threatened by the construction and
operation of the proposed plants" (ibid., page 3);
the Sierra Club intervened because it was "opposed to
the exploitation of such resources contrary to environmental enactments of federal and local governments"
(ibid., page 4); the United Auto Workers intervened
because it was concerned about the impact of radioactive steam on the workers employed in the area, and
about their general safety; Trout Unlimited joined to
preserve and ameliorate the trout areas of Michigan,
one of which was the Tittabawassee River; the West
Michigan Environmental Action Council intervened over
a concern about nuclear safety; and, finally, the
University of Michigan Environmental Law Society joined
in the petition to intervene to "demonstrate to the
people of the United States and Michigan, particularly
the young activist community, that environmental protection can be achieved through existing legal channels" (ibid., page 8).

The Mapleton Intervenors ("Mapleton") was the
only intervenor group openly opposed to the plant in
its entirety. It went on record as being committed to
obstructing the proceedings and hopefully, to delaying
the plant forever.

The hearings progressed slowly. In the late spring
of 1971, questions were raised about the safety of the

emergency core-cooling system (ECCS). In July, the courts ruled favourably in the "Calvert Cliffs" case, which promised to expand the issues beyond their present scope to new environmental controversies.

Meanwhile, public attention had turned to the delay. The substance of the hearings was, for many, uninteresting : Michigan had been and was pro-nuclear. The delay was blamed entirely on the "meanderings of a few isolated individuals". The news media challenged the right of the environmentalists to obstruct the goals of CPC. An article was published on the front page of the Midland Daily News under the headline : "An Open Letter to the Public; Will a Few Destroy Our Area ?" Throughout early 1971, the AEC received numerous letters from citizens and a few congressmen protesting delays in the granting of the construction permit (Ebbin and Kasper, 1972, page 72).

The Midland County Board of Supervisors voted unanimously to accept a resolution "that sets as the Board's goal final completion of an atomic energy plant at Midland" (Midland Daily News, 3rd September 1971, page 1). They even appropriated $20,000 to see the goal accomplished.

While the proceedings continued to receive "bad press", the arguments continued to rage. The lack of familiarity with the issues and the limited time provided to develop a case presented another problem in the proceedings. Generally, the intervenor must prepare a case within three months against a group that has been familiar with the material for at least two years.

Throughout the trial, the ASLB persistently refused to consider many of the requests for information filed by the intervenors. It repeatedly stated that there had been no showing of "relevance or need" and that "disclosure of the presently withheld documents would, in our opinion, be contrary to the public interest" (memorandum dated 3rd September, 1971). It said that internal papers "which may be of little or no significance" could produce "adverse effects on the regulatory process" (ibid.). This argument is not without truth. An excessive amount of material and testimony had been collected, and the time needed simply to read the documents was enormous. However, this continued to provoke the attorneys struggling to build a case.

By September 1971, debate had shifted to the environmental issues, in anticipation of the EIS. The handling of procedures again was the subject of much discussion. The intervenors disputed the ASLB's ruling that all environmental areas of controversy must be stated ahead of time.

On 7th January, 1972, the Staff of the AEC issued the draft EIS and circulated it for comment. The final EIS was made available in March, and the evidentiary hearings were scheduled to open in May.

Before the hearings, EDF formally withdrew from the proceedings after examining its other legal commitments, citing the qualifications of the other intervening groups and itsconfidence in them. (EDF was preparing for the ECCS generic hearings scheduled to begin soon in Bethesda, Maryland.)

On 17th December, 1972, the ASLB filed an initial decision that recommended authorisation of a construction permit. In accordance with Rules of Procedure, appeals were accepted. On 7th January, 1973, Saginaw moved to "revoke and recall the initial decision" because the board had been biased. The motion was denied.

Concurrently with the latter motion, Saginaw filed another with the Appeal Board.

In March, the Appeal Board issued an interim judgement on questions of quality assurance and quality control. It modified the initial decision to ensure that construction and implementation of the project would be adequately done by CPC and Bechtel Corporation, the architecture-engineering firm. It requested that a complete programme be outlined and filed, detailing actions to assure quality.

In the fall of 1973, the AEC declared a major change of policy on the conservation of energy. In the "Niagara-Mohawk Power" decision, the AEC held that certain energy-conservation issues must be considered during the environmental review. Soon after this decision, Saginaw sought clarification of the Midland Case in light of "Niagara". The AEC did not respond to the request until it was forced to do so by the United States Court of Appeals. When it did respond, the AED said, in refusing to reopen the proceedings (transcripts of "Aeschliman et al. versus United States Nuclear Regulatory Commission" and of "Saginaw Valley Nuclear Study Group versus United States Nuclear Regulatory Commission") * :

"We will not apply Niagara retroactively to cases which had progressed to final order and issuance of construction permits before Niagara was decided."

---

* "Nelson Aeschliman et al. versus United States Regulatory Commission," United States Court of Appeals, Docket Number 73-1776, 1973; and "Saginaw Valley Nuclear Study Group versus United States Nuclear Regulatory Commission," United States Court of Appeals, Docket Number 73-1867, 1973.

In February, Dow and CPC renegotiated their contract. The renegotiation had radically altered the commitment Dow had with regard to the reactor. Dow cut
its purchase of steam by half, to only 900,000 kilograms
of steam per hour and, although they might, they were
no longer required to purchase electricity from the
plant. Saginaw again requested the AEC to reopen the
hearings, stating that with the CPC-Dow contract thus
changed, the original benefit-cost analysis was no
longer accurate. Saginaw quoted from the original EIS
in support of its argument.

However, the AEC repeatedly declined to review
the case saying, in part :

"Dow still intends to (take) substantial takes
of electricity and steam and intended to maintain
their fossil-fueled facilities 'primarily on a
standby basis'."

Saginaw had held that the EIS specifically stated
that Dow's fossil fuel-powered facilities would be
closed entirely. Therefore, in Saginaw's understanding,
this sufficiently altered the benefit-cost ratio.

Throughout the requests of Saginaw, the AEC maintained its original rejection in saying that before a
licensing board must explore conservation alternatives,
intervenors must "state clearly and reasonably specific energy conservation contentions in a timely fashion".

Rebuked enough by the AEC, Saginaw took its contentions to the United States Court of Appeals. In
November 1974, they argued their case. In April 1975,
the Court put the case in abeyance until a related
case ("NRDC versus NRC") was resolved. Almost two years
later, in July 1976, the court ruled on the case and
upheld almost every contention raised by the Saginaw
intervenors. In so ruling, the license was remanded
for further proceedings. The first argument that Saginaw had contended was that the EIS, especially in
view of the CDC-Dow renegotiation, did not sufficiently
consider alternative sources of energy, the most important of which was conservation. In deciding, the Court
of Appeals referred to an earlier decision, the "Calvert
Cliffs" case. In that case, the court had held that
"it is unrealistic to assume that there will always
be an intervenor with the information, energy and money required to investigate environmental issues"
(transcripts of "Aeschliman et al. versus United States
Regulatory Commission" and of Saginaw Valley Nuclear
Study Group versus United States Nuclear Regulatory
Commission"). Primary responsibility, therefore, lay
with the AEC to fulfill the requirements of NEPA. The
AEC had erred, the court held, in "promulgating a
'threshold rest' which essentially requires the intervenor to prove an alternative satisfies the 'rules of

reason' before the commission will investigate it" (ibid). The only responsibility of the intervenor is to bring "sufficient" attention to an alternative. Thereafter, it is the responsibility of the AEC to investigate it.

Conservation, the court held, had not been given sufficient attention in the EIS. In requiring a re-opening of the proceedings, the court requested the ASLB to ensure that conservation is "explicitly" con-sidered and to "take into account the changed circum-stances regarding Dow's need for process steam, and the intended continued operation of Dow's fossil-fueled generating facilities" (ibid., page 21.).

In the meantime, CPC is threatened with bank-ruptcy. Costs on the Midland plant have skyrocketed from an original $554 million to over $1.9 billion (Detroit Free Press). CPC has threatened to sue Dow should it back out; Dow in its turn has threatened to countersue, while it has weakly stated its continued commitment to the plant.

c)   Analysis and Conclusions

The Midland Case has been one of the most pro-tracted disputes over a nuclear powerplant in the United States. Original estimates had called for the plant to be operating by 1974, but now optimistic estimates are that it will be operating by 1982 - assuming it is built at all. The hearings that are now reconvened may put an end to the plant in its entirety, regardless of the fact that the plant is one-fifth completed, the cooling pond is finished, and over $600 million have been spent.

This raises several questions regarding the "pro-priety" of the intervenors and the format within which they are forced to intervene. Whether or not the power-plant at Midland will be completed is not the question. Instead, the situation calls for an examination of the procedural inequities and inherent problems that h a v e   b e e n besetting the Midland controversy.

The Midland Case has become a war of attrition, each side waiting for the other to concede defeat. Both sides are close to doing so. CPC is close to bankruptcy, unable to find sufficient funding to com-plete its efforts to reach the atomic age. The inter-venors are no better off : Myron Cherry, attorney for Saginaw, is owed over $23,000, and was recently compel-led to drop out of the hearings because an anonymous sponsor has run out of funds, and Mary Sinclair has reached the end of her tether, after having raised over $150,000.

Neither side had intended the conflict. The intervenors, excluding the Mapleton Intervenors - publicly at least - espoused goals other than the cancellation of the plant through delayed hearings. (Now, however, they are committed to stopping the plant, their financial resources permitting.)

An attitude of noncompromise pervades the Midland controversy. The staff and the Applicant made no attempt to facilitate the efforts of the environmental groups. In their turn, the environmental groups caused problems and extended delays. What was created is what one author calls a "no-win" situation (Irving Like, personal communication). Neither side ever wins as the vast and complex process continues. The outcome is unsuccessful for either contender in this dispute; hence, the procedure itself must be called into question.

The role of environmental groups in the Midland Case took one predominant form - intervention. As intervenors, they sought, through the licensing process, to raise any contentions and generally argue against the plant proposal. Other functions, such as the public's increasing knowledge about the case, were subordinate. This has had serious consequences at Midland. The environmentalists have not been successful in gaining a broad base of public support, like they did in Bodega. Instead, their energy and focus has been in the legal realm, through the licensing agency and the courts.

Any party who might be affected by the outcome of the proceedings may intervene. This is a broad basis for intervention. However, it becomes further specified by requiring the intervenor to declare certain areas of affect, or concern. This must be accompanied by a showing of relevance. The intervenors in the Midland Case were motivated to intervene by several concerns over social, environmental, and technical problems.

Motivation for intervention, regardless of the stated goals, was prompted by a desire for more public input in the decision-making process. The Midland intervenors were demanding the right for a more assiduous review of the promises and assurances that had been blanketly given by the Utility and the Commission staff. This was a desire on the part of the intervenors to make the government more accountable, and the public more involved in, the environmental and technological assessment of the proposed facility. There was a time not long ago when assessment of the effects of technology were performed by experts based upon narrow considerations of technical efficacy and direct observable results. But the growing awareness of the pervasiveness of the impacts of technology has led to the demand, on the part of the environmental groups through

69

the intervention process, of the inclusion in the assessment process of the indirect, less tangible, social and environmental effects that often accompany new technology (Ebbin and Kasper, 1974).

Scepticism about the ability of 'experts' to make decisions and to anticipate all technological impacts of the plant underlies much of the environmentalistic desire for involvement in the decision-making process. Midland, in this respect, is paradoxical. Most of the town of Midland had accepted the assurances of Dow's and Consumers' scientists that the plant would be safe (Detroit Free Press, 28th June, 1971). Yet Mary Sinclair and the Saginaw Nuclear Study Group were not so persuaded. The group was concerned, as were other groups, about leaving the choices up to the experts alone. In testimony before the Michigan State Water Resources Commission in 1970, Mary Sinclair reiterated this point of view when she said :

"The most compelling argument for the value for the rise of public participation in the decision-making process is the fact that by now it is clear that citizens have for too long left the decisions in energy development to the so-called experts - to the corporations, commercial organisations and government agencies who develop powerful vested interests in certain decisions. The fact that these decisions were not adequately assessed by public scrutiny is one of the reasons we find ourselves in deep trouble in many aspects of our economic, social and political development today."

Technological assessment as one of the professed goals of the intervenors was successful in the Midland Case by upgrading the safety and polluting effects of the proposed plant. This took two forms : delaying construction until certain safety features were upgraded and assuring that the plant would be made as safe as possible under existing technological capabilities. The upgrading and redesigning of certain plant functions was accomplished by the groups' scrutiny *. Furthermore, quality assurance and safety evaluations were insured by the intervenors.

However, upgrading of the specific plant at Midland was but a portion of the lengthy hearings. There were several reasons why scientific arguments were not

---

* This has been one of the most successful aspects of environmental intervention. In another case (Palisades), the environmentalists predominated and forced the utility to install cooling towers, a closed-cycle system, rather than dump the waste heat into the Great Lake Michigan.

taken up in the Midland Case. Most important was the fact that CPC had confronted the lawyer, Myron Cherry, a year before in the Palisades Case, and was prepared for many of the arguments he presented. Thus, the scientific disputes that might have arisen in the Midland Case had already been resolved. CPC's concessions in the Palisades Case had cost more than $51 million, a fact that forced CPC to come prepared to deal with Cherry.

Another reason why scientific questions were not raised in the Midland Case is that the ECCS hearings had removed discussion of the ECCS from the trial. This could have had far-reaching consequences, since the information that Myron Cherry and the Union of Concerned Scientists brought out at the ECCS hearings about the AEC's suppression of data on the question of nuclear-reactor safety and the ECCS was devastating.

During the course of the trial, the AEC responded to the repitition of issues raised by each individual siting case, by making them the concern of generic rulemaking hearings, thus divorcing those issues from a particular site. This was done in 1971 with the emission standards for radioactivity and again in 1972 for the ECCS. The bulk of the hearings in the Midland Case were consumed, and concerned with, legal wrangling, making it difficult to ensure that the scientific and technical aspects of the case were considered.

This has several implications. The hearings are designed on an adversarial platform. The intervenors may therefore approach the issues by offering counter scientific testimony or challenging the scientific testimony given. In challenging other testimony, they are really questioning whether procedural requirements have been fulfilled - in other words, whether the conclusions had been substantiated by proper information and sufficient evidence. This is really their only alternative : ascertaining and ensuring that statutory and procedural requirements have been fulfilled and whether the conclusions have been rationally substantiated.

There were problems involved in the Midland Case for the intervenors when they tried to find and provide scientific testimony apart from the problems involved in subpoening witnesses of the applicant. One problem is the lack of sufficient funds. This required the intervenors to engage the advice of graduate students and other scientific representatives whose services were within their budget. This testimony is often regarded by the ASLB as less substantial.

Another problem is the one addressed by Myron Cherry in his letter to the Chairman. He wrote (5th May, 1971) :

71

"I am sure the Board is aware that opponents of a
construction permit for a nuclear plant cannot
easily find witnesses or indeed assistance from
the ranks of industry or academicians who for one
reason or another are so geared to pursuing their
professions that they cannot render objective
assistance to intervenors. Accordingly, intervenors must, not only out of preparation, but of
necessity, channel their preparation of their
case into securing information from unfriendly
sources... For this Board to render decisions
which destroy intervenors' rights with respect to
adequate pre-trial preparation by way of inspec-
tion of documents and interrogatories is, in the
light of the circumstances, tantamount to denying
the intervenors the right to prepare and hence
participate."

Finding scientists capable of providing their
time, regardless of their fee, is difficult. Therefore, to build any case at all the intervenors are
required to utilise the crucible of cross-examination.

Finally, the problem of the expert leads to the
general discussion of the "propriety" of expert advice. The major assertion of the intervenors is that
the technological decisions are too interrelated for
a decision by one, or another, expert. Rather they
assert the need for a social scientific assessment more
appropriate to a judicial or inter-disciplinary board,
not an atomic regulatory committee.

Undoubtedly, scientific testimony and its problems
plays an integral part in the proceedings. Yet more
important is the ability of the legal representative.
Since the discussion centers on the procedural mechanism and statutory terms and their fulfillment, the
legal aspects, and thus the lawyer, become crucial to
the hearings.

This gives the pivotal role in the proceedings to
the lawyers. The main protagonists in the Midland Case
were the environmental groups' lawyers, Myron Cherry
and Anthony Roisman. This might well represent a pivotal point in anti-nuclear disputes; the replacement
of echnicians with lawyers (Herrera, 1971). Since the
success of intervention depends on the enforcement of
procedure, legality becomes more important than the
technical questions involved.

The Midland Case has been described as a case of
all substance and no strategy (Ebbin and Kasper, 1974).
Because of the lawyers, predominance, important testimony is often overlooked, or misleading information
is given, due to their lack of expertise. Both Anthony
Roisman and Myron Cherry admit to their unfamiliarity

with the issues. To compensate for this Myron Cherry often has a technical advisor by his side.

Many people feel that the intervenors try to test the agency process rather than questions of technical fact. But this is not inconsequential. Fulfillment of the procedural and administrative functions of the agency is important. If these are not fulfilled, then it provides ground for an appeal decision later. The pivotal concern of the environmental groups is to see that the requirements specified by NEPA and the Atomic Energy Act are fulfilled. In the Midland Case, the Court of Appeals remanded the proceedings because the AEC had failed to properly complete the ACRS report and had also failed to consider alternatives to the facility, especially conservation.

This raises questions about the propriety of an administrative review board to make decisions of this scope. The hearings were originally designed to allow for public input and to raise and review scientific and technological issues. By allowing the public to part-icipate, it was felt that their fears could be assuaged and the general safety and quality of the facility appraised by the public at large. It is doubtful that one hearing could accomplish all these goals, let alone accomplish a de novo review of the environmental impact of the proposal.

The problem of measuring social costs is more extreme in the Midland Case than in the Bodega Case. Ostensibly, they have a statutory requirement to ful-fill, under NEPA. Understood in broad terms, this pol-icy requires a complete assessment of the environmental consequences, including within it such areas as site and design alternatives. The ASLB, a board whose mem-bers are chosen for their administrative and perhaps their nuclear expertise, is asked to consider terms beyond its area of expertise. Even the terms them-selves, the social costs are, as they were in Bodega, intangible and rarely quantifiable. Furthermore, the judicial system is continually expanding the scope and depth of their consideration.

The Board is required to make certain conclusions based upon the testimony provided during the course of the hearings. The Board is allowed and expected to use its expert knowledge and experience in evaluating and drawing conclusions from the evidence in the record (CFR 10, Part 2). The decision must rest upon the com-plete evaluation of all aspects of the testimony as well, including the social benefits and costs of the proposed plant.

In the Midland Case, the ASLB and the Commission refused to discuss issues like conservation, which they felt were beyond their scope as a review board.

73

The Court of Appeals reversed this decision by stating that it was part of their responsibility. Is a review board, at the end of a licensing process, the proper place for scientific discussion and evaluation of problems such as environmental safety, impact and alternatives ? Furthermore is the adversarial type of hearings the proper method for discussion ? The Appeals Court's decision expanded the scope of the commission's and the ASLB's responsibility. But are they qualified to decide upon this and other issues when they are contested ?

This leads to another problem raised by the process. The hearings are designed to give the public a chance to participate, and yet they occur at the end of a long process of prereview and discussion between the staff of the AEC and the applicant. By the time of the hearings, the staff of AEC had already determined that the plant would be safe and the environmental costs not greater than the benefits. They approach the hearings with the plant a foregone conclusion, in their minds. The ASLB, functioning as an independent part of the larger commission, must feel a certain bias in that regard. Arguments that are raised, or areas of concern that are mentioned, might well have been fully discussed already, beforehand. Therefore, rediscussing them is often fruitless.

The consequence of these problems is delay which, depending upon one's point of view, is either necessary or quite unnecessary. Expansion of the scope of the AEC's responsibility and the increased involvement of environmentalists in the proceedings did much to contribute to the delay in the Midland Case. The costs of this delay have already been commented upon. The Midland Case demonstrates the need for a change in the administrative method, so that the decision-making process can be improved.

The impact that environmental groups had in the Midland Case can be divided into three categories : modification of the proposed plant's design, insisting that conservation be considered as an alternative to the project as proposed, and education of the public. In terms of modification of the proposed facility, the environmental groups have been successful. The CPC was of course reluctant to accept much modification. But in the Midland Case, as in other nuclear disputes, the environmentalists were most successful in this area. Some of the modifications instituted were changes in the intake pipe, the relocation of several "getaway" transmissions lines, and a reduction in the emission of radioactive materials.

Another type of modification in the proposed project was the requirement for surveillance. Since the

ASLB is required to decide on the safety of the plant,
the licensing decision will often be contingent
upon proper surveillance of certain areas by the utility. This was agreed to by the CPC in the Midland Case.
CPC promised to regulate phosphate levels in the river,
to ensure that radiological standards were enforced,
to provide ecological studies, and to guarantee,
through surveillance, proper and safe construction.
These points originated in the environmentalists'
arguments.

As for the introduction of conservation as an
alternative approach, or whether or not the plant
should be built, the answer is yet to come. The environmentalists offered a strong challenge at Midland
for a "no-growth" policy. So far, they have been successful in guaranteeing that this issue will be more
thoroughly considered in the reconvened hearings.

The efficacy of the environmentalists in educating
the public is difficult to assess. However, one thing
is certain : they did succeed in raising sufficient
doubt about the project. This is coupled with the economic doubt on feasibility that now surrounds the project as well. How much public scepticism has been increased is difficult to measure because other factors,
such as the economics issue, have been involved.

6. Abbreviations, Acronyms, and Short Titles

AAAS	American Association for the Advancement of Science
ACRS	Advisory Committee on Reactor Safeguards
AEC	Atomic Energy Commission
AINWS	Arctic International Wildlife Range Society
ANWR	Arctic National Wildlife Range, Alaska
APA	Administrative Procedures Act
API	American Petroleum Institute
ASLB	Atomic Safety and Licensing Board
ASLAB-ALAB	Atomic Safety and Licensing Appeal Board
Brookhaven Report	See : "WASH-740"
CAPBHH	California Association to Preserve Bodega Head and Harbor
Central Hudson	Central Hudson Gas and Electric Company
CEQ	Council on Environmental Quality
CFR	Code of Federal Regulations

CNI	Consolidated National Intervenors
Con Ed	Consolidated Edison Company of New York
Corps	Army Corps of Engineers, Department of Defense
CPC	Consumers Power Company
CPNC	Certificate of Public Necessity and Convenience
CPUC	California Public Utilities Commission
CZM Act	Coastal Zone Management Act of 1972
DOI	United States Department of the Interior
DOT	United States Department of Transportation
DWP	Deepwater port
DWT	Deadweight ton
ECCS	Emergency core-cooling system
EDF	Environmental Defense Fund, Incorporated
EIS	Environmental impact statement
EPA	United States Environmental Protection Agency
EPC	Environmental Policy Center, Washington, D.C.
ERCA	Energy Research and Development Agency
FEO	Federal Energy Office
FOE	Friends of the Earth, Incorporated
FOI	Freedom of Information Act
FPC	Federal Power Commission
LOOP	Louisiana offshore oil port
MMF	Committee on Merchant Marine and Fisheries, United States House of Representatives
MW	Megawatt
NECNP	New England Coalition on Nuclear Pollution
NEPA	The National Environmental Policy Act of 1969
NPCA	National Parks and Conservation Association
NPS	United States National Park Service
NRC	Nuclear Regulatory Commission
NRDC	Natural Resources Defense Council
NWF	National Wildlife Federation
Pet 4	Naval Petroleum Reserve Number Four, Alaska

PG & E	Pacific Gas and Electric Company
PI	Project Independence
PSAR	Preliminary safety analysis report
PUC	Public Works Committee
PWC	Public Works Committee, United States House of Representatives
SAPL	Seacoast Anti-Pollution League
SC	Sierra Club
Scenic Hudson	Scenic Hudson Preservation Conference
SCLDF	Sierra Club Legal Defense Fund
SPM	Single-point mooring
TAPS	Trans-Alaska Pipeline System
TNC	The Nature Conservancy
Trail Conference	New York - New Jersey Trail Conference
TWS	The Wilderness Society
UCB	University of California Berkeley
UCS	The Union of Concerned Scientists
VLCC	Very large crude carriers
"WASH-740"	"The Brookhaven Report" (full title: Theoretical Possibilities and Consequences of Major Accidents in Large Nuclear Power Plants)
WEST	Western Energy Supply and Transmission

7. References

Anderson, Frederich R., 1973. NEPA in the Courts. Resources for the Future, Washington, D.C. 374 pages.

Culhane, Paul J., 1974. Federal agency organisational change in response to environmentalism. Humboldt Journal of Social Relations 2 : 31-44.

Ebbin, Steven and Raphael Kasper, 1974. Citizen Groups and the Nuclear Power Controversy : Uses of Scientific and Technological Information. The MIT Press, Cambridge, Massachusetts. 307 pages.

Friesema, Paul H. and Paul J. Culhane, 1976. Social impacts, politics, and the environmental impact statement process. Natural Resources Journal 16(2) : 339-356.

Messing, Marc and David Rosen, 1972. Popular Participation in Environmental Planning in the United States : A Case Study of Earth Day and Environmentalism as a Popular Social Movement in the United States in 1970. (Commissioned by the United States Department of Economic and Social Affairs, Social Development Division, in preparation for the 1972 Stockholm Conference on Man and the Human Environment.) Friends of the Earth, New York. 97 pages (unpublished manuscript).

O'Riordan, Timothy, 1976. Policy making and environmental management : Some thoughts on processes and research. Natural Resources Journal 16(1) : 55-72.

Sewell, W.R. Derrick, and Timothy O'Riordan, 1976. The culture of participation in environmental decision making. Natural Resources Journal 16(1) : 1-22.

Sive, David. 1973. The role of litigation in environmental policy : The power plant siting problem. Pages 65 to 75 in : Albert E. Utton and Daniel H. Henning, editors, Environmental Policy: Concepts and International Implications. Praeger Publishers, New York. 266 pages.

Wengert, Norman. 1976. Citizen participation : Practice in search of a theory. Natural Resources Journal 16(1) : 23-40.

CHAPTER III

DEVELOPMENTS IN THE PROCEDURES
FOR SITING THERMAL POWER PLANTS IN FRANCE

1. Introduction

    Since the French energy policy leaders have opted
in favour of the development of nuclear power, it can
be assumed that in future, among the problems raised
by the siting of major energy facilities, those concerning nuclear power plants will be in the limelight
in France.

    Accordingly, this paper sets out the current pro-
cedure for siting nuclear power plants (more generally,
both conventional and nuclear thermal power plants),
as it emerges from a number of changes in previous
years.

    It might also be added that the sole aim of this
paper is to report on recent developments in the siting procedure . More specifically, it does not include information which would enable these developments
to be interpreted within the context of opposition
movements, although the latter have undoubtedly favoured
these changes. This would require detailed analysis of
nuclear protests as they have occurred in France and
in particular the links between this overall phenomenon and local opposition to sites.

    As such an analysis remains outside the scope of
this paper, any attempt at assessing the extent to
which recent developments in the siting procedure are
likely to reduce local opposition in future would be
hazardous.

    The siting procedure can be presented in the form
of three main successive stages, although they overlap
in many cases :

    -   the first stage consists in the determination
        and preliminary selection of sites by the cen-
        tral government departments in conjunction
        with the regional and departement authorities;

    -   the next consists in discussing and examining
        the preliminary selection with elected repre-
        sentatives and local communities at regional,

département and municipal levels with a view
to reaching agreement on one or more sites.
During this examination, many preliminary
studies are drawn up, including the environmental impact assessment;

- the third concerns the application procedure
for a "Déclaration d'Utilité Publique" (DUP -
statement that the project is in the public
interest) on the construction of the power
plant at one of the sites agreed at the previous stage. This stage includes a "public
enquiry" and the publication of the "environmental impact assessment" referred to above.
At the end of the stage, the DUP decree is
signed, confirming "official recognition by
the State of the advisability of the proposed
operation" *.

This paper is divided into three sections because
they correspond to the three stages referred to above.
They will be considered in detail below.

2. Determination and preliminary selection of sites
by the government departments

In France, the responsibility for determining
suitable sites for the construction of power plants
is held by Electricité de France (EDF), the national
electricity supply body, within the framework of its
general functions of power generation, transport and
distribution **. The first step in the determination
involves the search for a priori "possible" sites.
This is done mainly with the support of maps and local
documents, with site visits being restricted to a minimum. At the end of the operation EDF has a list of
"possible" sites, although they are not necessarily
"feasible" from the technical and economic points of
view. The next step is to propose one or more of the
"possible" sites to which EDF would like access for a
fixed period of time in the light of its regional
consumption forecast, for government department examination at inter-ministerial meetings.

A few of these sites are then selected, each government department dealing with the criteria relevant
to its own sphere. Thus, the meetings are generally

---

* This is how the DUP is currently presented by the
official authorities.
** Defined by the Gas and Electricity Nationalisation Act
of 8th April, 1946 and its implementing decrees.

attended by : DATAR *, EDF, DIGEC **, SCSIN ***, the
Ministry of the Quality of Life, SCPRI ****, DAFU *****
and other bodies if necessary.

The criteria put forward by each of these government departments   in order   to reconcile the following
interests during the examination of the sites should
be mentioned:

- balance of regional power generation and consumption and structure of the high-tension
  network according to EDF forecasts for the
  end of the century;

- safety of the facilities designed by SCSIN
  from two angles : the way in which the outer
  environment can affect plant safety, and con-
  versely the way in which the facility can affect the environment, and of course, the surrounding population, in the event of an acci-
  dent;

- relationships between the general features of
  power plant sites and radiation risk implic-
  cations (question concerning mainly SCPRI);

- multiple environmental impacts and their dependence on the siting and installed capacity
  on the site ****** (question dealt with by the
  Ministry of the Quality of Life);

- general regional planning trends with, among
  other aims, that of reducing major imbalances
  between already industrialised regions and
  those likely to become so by the end of the
  century, between the Paris region and the pro-
  vinces and between rural and urban zones.

---

    * Délégation à l'Aménagement du Territoire et à l'Action
Régionale, (Regional Planning Committee), currently attached to
the Ministry of Equipment and Regional Planning.
    ** La Direction du Gaz, de l'Electricité et du Charbon au
Ministère de l'Industrie (Gas, Electricity and Coal Directorate,
Ministry of Industry).
    *** Le Service Central de Sureté des Installations Nuclé-
aires (Central Service Department for the Safety of Nuclear Installations), Ministry of Industry.
    **** Le Service Central de Protection contre les Rayonnements
Ionisants (Central Service for Protection against Ionizing Ra-
diations), Ministry of Health.
    ***** La Direction de l'Aménagement Foncier et de l'Urbanisme
(Directorate for Building Planning and Urbanisation), Ministry
of Equipment.
    ****** The sites required in France should be able to hold at
least four 900 Mw reactors.

For instance, DATAR wishes to give priority to the industrial development of western and south-western France;

- the numerous indirect implications of the site choice, i.e. both those concerning infrastructures associated with the site and those in terms of urbanism, population patterns, creation of local employment, pollution related to the position of the site in the landscape, etc. These somewhat indirect considerations concern DAFU.

In some cases, these meetings at central government level may be supplemented by meetings of the regional or département prefectures(s) concerned if the sites meet the above criteria. One of the purposes of these internal administrative meetings (meetings attended by the elected representatives are held at a later stage), is a preliminary examination of the local population's likely response to the project(s) and more specifically to the role which might be played by the elected representatives, trade associations, environmental protection associations etc.

Such inter-administrative consultation, which is aimed at combining the interests of many ministerial departments at the beginning of the procedure for siting thermal power plants, constitutes an improvement over previous practice. In the 1950s and 60s, the siting procedure conducted by the civil service was controlled solely by the Ministry of Industry in close liaison with EDF, apart from a few exceptions *.

To a certain extent, the requirements of modern economic development in France in the early 1970s meant that the nature of the criteria applied by the EDF and the State, which has remained unchanged during the previous 20 years, was called into question.

In accordance with the targets fixed in the 1960s with a view to reducing regional economic imbalances, the new economic development measures entailed the intervention of the regional planning government department in the procedure in order to introduce new criteria. Furthermore, the State's increasing concern for the environment, which first became apparent at the end of the 1960s and resulted in the creation of a Ministry of Environment, led to new regional planning concepts integrating all the "external" features hitherto disregarded in the planning process. It should be noted that these new criteria began to be taken

_____

* An example is the inter-ministerial consultation held aroung 1955 by the Service d'Aménagement de la Région Parisienne (Paris Region Planning Department) when two sites were selected out of 20 submitted by EDF.

into account before it was decided to develop a major nuclear power programme *. Nevertheless, technical developments in the generation process and in particular the increase in unit capicity and the switch-over from conventional thermal generation to nuclear generation ** have probably favoured the revision of the siting concepts for these industrial units, which used to give precedence to technical and economic criteria only ***.

In practice, the new situation has been reflected by the gradual establishment of the inter-ministerial consultation procedure described above. It should be pointed out that this procedure has enabled 34 sites to be selected from about 100 "possible" sites proposed by EDF, and these have been the subject of the "Regional Consultation" announced in November 1974 (to be described in greater detail in the section describing the second stage of the siting procedure). The current procedure (i.e. that applied since early 1977) consequently concerns the generation of sites intended to replace some of those on which a Regional Consultation had been held**** and those required for meeting long-term requirements.

---

* Thus, well before the Government's March 1974 decision ("Messmer Plan") to launch a major nuclear power programme, certain proposals for traditional thermal power plants had been the subject of reservations by the Ministry of Environment owing to air pollution problems in industrial areas. As a result of this concern, the Comité Interministériel d'Actions pour la Nature et l'Environnement (CIANE - Interministerial Action Committee for Nature and the Environment) decided on 20th July, 1972 to make the Ministry of Industry responsible for siting oil refineries and conventional thermal and nuclear power plants.

** The increase in unit size raises the problem of the ability of the environment, in its broadest sense (i.e. ecological, health, psycho-sociological etc.) to "absorb" the facilities. Furthermore, the switch-over to nuclear power considerably reduces the constraint of siting the facilities close to fuel sources.

*** Some proposals for nuclear power plants have also been the subject of reservations by certain government departments, e.g. the extension of the Saint Laurent des Eaux power plant and the construction of the Dampierre-en-Burly plant on the Loire, whose DUP applications filed in October 1971 were criticised by the government departments of the Environment, Cultural Affairs and Equipment. The latter wished to preserve the Loire Valley's role of "Métropole Jardin" (a region to be developed while preserving its recreational assets). The Gravelines site, Nord, raised a similar problem for DATAR, as, in the latter's opinion, it affected the "green lung" planned between the Dunkirk industrial belt and the Gravelines residential belt.

**** This consisted in enabling Regional Councils to give their opinions on the selection of a dozen sites out of the 34 which would hold the facilities needed to meet power requirements until 1988.

In short, at the end of this stage, the government has a list of "possible" sites agreed to a priori by all ministerial departments.

The second stage consists in consulting the elected representatives and local authorities in order to prune the initial selection and at the same time to conduct so-called "preliminary" studies aimed at assessing the technical and economic "feasibility" of the project and its acceptability in terms of safety, radiation protection * and environmental impact.

3. Discussion of the preliminary selection at regional and local government level; preparation of preliminary studies

Once the Regional (or Département) Prefect has been instructed by the government to start the Regional (and/or Département) Council consultation stage, aspects of the project or projects concerning the Region or Département are then discussed.

The discussion obviously covers all regional and local implications of the siting of one or more power plants, at economic, financial, population (including construction labour problems), agricultural, ecological etc. levels. Government and EDF experts intervene in the discussions, dealing in detail with the various aspects referred to above, and in some cases representatives of nature-protection associations which are sufficiently influential and reputable to be regarded as "valid" spokesmen are invited to present their points of view.

These discussions cannot be considered as preliminary work towards a decisive vote on the future of the project(s) examined. Neither the Regional Councils nor the General Council ** hold any legal right to restrict the government's freedom of action regarding site selection ***.

In fact, these discussions are considered by the government as a suitable means of consulting regional and local bodies with a view to taking their preferences

---

* Without going into detail here, it should be pointed out that the safety and radiation protection studies referred to are not the detailed studies undertaken during the preparation of the Building Permit Application, which is independent of the siting procedure in question. The studies referred to here are those concerning safety and radiation aspects relating to the specific features of each site and which may result in the site being rejected.

** In other words, the Département Council.

*** A brief description of the composition and functions of Regional Councils is annexed hereto.

into account in the procedure leading to final site selection. Consequently, they act as a means of obtaining, or at any rate assessing, the acceptability of the project to the regional and local official representative bodies *.

During the Regional Council consultation stage, the EDF carries out the various preliminary studies aimed at assessing the "feasibility" of the project or projects from the technical, economic, safety, radiation protection and environmental angles **. Consequently, they may lead to the decision to turn down a site if the latter does not meet the required criteria ***. They take one to two years to complete.

Councils are kept informed of progress in the studies, which are also included on the agendas for inter-ministerial meetings every time that new results call for a decision. At such meetings, any aspects which are giving rise to problems for certain government departments are reviewed. The aim is to reach agreement of all parties present on the "feasibility" of the site in the widest sense, in other words not only under criteria specific to EDF, which are mainly technical and economic, but also from a safety, radiation protection, environmental etc. angle.

---

* It should be pointed out that the conditions for acceptance of a project by the Councils differ widely according to the regions and with time. For many reasons which need not be dealt with here, the Councils' attitudes towards the nuclear programme have appreciably changed since the launching of the Regional Consultation procedure in January 1975. Very early on in the application of the latter, it became apparent that several regions were in favour of sites. Thus, at least five out of the 21 regions consulted approved some of the proposals concerning their territory as early as January or February 1975, while others waited until the end of the year before deciding, and one of them has not yet given its opinion.

** Although most of the studies are conducted in EDF laboratories and EDF consultancy bureaux, some are nevertheless subcontracted to outside bodies : most of the environmental studies are thus entrusted to the Commissariat à l'Energie Atomique (CEA - Atomic Energy Commission), the Institut Scientifique et Technique des Pêches Maritimes (ISTPM - Sea fishing, Scientific and Technical Institute), the Centre National d'Exploitation des Océans (CNEXO - National Ocean Exploitation Centre), the Centre d'Etudes Techniques du Génie Rural des Eaux et Forêts (CETEGREF - Water and Forestry Rural Engineering Technical Study Centre) and universities or regional laboratories, depending on the problem.

*** This happened in the case of the Corsept site, which was one of the 34 sites submitted for consultation.

Depending on the sites, several cases may arise :

- acceptance by the Councils may not in fact be based on the outcome of the studies and may sometimes be decided before completion of the studies *. The examination of the latter by the government departments is therefore relatively independent of the first procedure;

- in other cases, the Councils are reluctant for a number of reasons, including the many uncertainties on the potential risks of the project or projects. Consequently, they subordinate their decisions to the outcome of further studies. This situation inevitably affects the progress of the inter-administrative consultation on these sites.

Insofar as the additional studies requested by the Councils ** generally concern aspects related to population safety and environmental protection, the government departments responsible for these sectors receive the benefit of this outside help during their discussions with the "promoter" government departments.

In terms of the relative power of the various institutions, the Ministry of the Quality of Life is the main beneficiary of this interaction between the two procedures ***. At the end of the two procedures, the public authorities are able to assess the acceptability of the project or projects proposed on the basis of the positions adopted by the Councils consulted and also through the opinion of the Regional or Départe-ment Prefect, which is of major importance ****.

---

* Thus for instance, the Flamanville and Saint-Maurice-l'Exil sites received a favourable vote in early 1975 before the environmental impact assessments had been completed.

** These are not in fact the only ones : in most cases, the waterboards responsible for water resources management and improvements to or maintenance of water quality also intervene. On occasion they may have to give qualified or even unfavourable opinions on certain sites and to request additional studies.

*** It was in the awareness of the advantages that the Ministry of the Environment could gain from decentralising the discussions on siting that certain high-ranking officials in the Ministry supported the regional consultation proposal as soon as it was put forward. This aspect will be referred to later in the context of the development in this stage of the siting procedure.

**** The Prefect's opinion is a determining factor in the decisions whether the central public authorities should continue the operations, insofar as he can give the government some idea of the extent of the subsequent response to be expected, in the light of his knowledge of the regional situation.

Furthermore, after examination of the preliminary studies at inter-administrative meetings, either the site is rejected or it is concluded that at the stage reached in the studies the site is "feasible" even though later studies might subsequently lead to changes in the design of the proposed facility.

Once the site has been deemed "feasible" by the government departments, consideration linked to regional and local acceptability as described above will finally induce the public authorities either to defer any new decisions or to authorise the EDF to file a DUP application before the Ministry of Industry.

Once a project has reached this stage, it can be regarded as having very good chances of success, in spite of the many difficulties which may arise during the examination of the DUP application, i.e. the third stage described in this paper.

Before embarking on this new section, we shall recall the differences between the siting procedure described above and that used several years earlier.

The trend towards increased inter-ministerial consultation was referred to in the context of the preliminary site selection. Obviously, the consultation occurs at each stage in the procedure. It is worth emphasizing, however, that an environmental impact assessment has been included among the preliminary studies required. The need for such an assessment had already been officially recognised in the government circular to the Regional Councils in December 1974 *. In July 1976 the environmental impact assessment became compulsory under a law ** covering all major equipment or construction work, including nuclear power plants.

Although the important implementing decree for Article 2 of the Act has not yet been published, both the nuclear power plant projects to which the 1974 consultation was applied and those of the subsequent generation include an environmental impact assessment in line with the instructions set out in an inter-ministerial paper dated May 1975 ***.

---

* In this paper, entitled "Siting of Nuclear Power Plants", one of the preliminary studies mentioned is the "Study of the Power Plant's Environmental Effects, Influence on Wild Life, Increase in Water Temperature, Sites and Landscapes, Noise, Climate", etc.
** Article 2 of the Act of 10th July, 1976 on the protection of nature. See Annex II for official texts.
*** A circular dated 24th August, 1976 from the Ministry of Industry states that "the impact assessment will be carried out on the basis of the list drawn up at inter-ministerial level on 7th May, 1975 and sent to the Prefects on 11th June, 1975 by the

To large extent the future role which the impact assessment might have to play in the siting procedure depends on the way in which it is promoted. As this is defined with reference to the filing of the DUP application, this matter will be considered as a whole in the third part of this paper.

The other noteworthy development is the way in which discussions between the government and regional and local parties will henceforth be conducted. Previously, relations with these bodies merely consisted in unofficial contacts, mostly on the part of the EDF representatives, in order to estimate the local acceptability of the project. Once the preliminary studies needed in order to assess the project's technical feasibility had been completed, the decision to continue the project was made by the Ministry of Industry upon a proposal by EDF. The project was then officially announced in the region concerned in liaison with the Prefect.

The examination of the DUP file could then begin. The period between initial unofficial regional contacts and the official announcement of the project varied according to the case but did not exceed one year.

At present, in other words since the establishment of Regional Consultation in January 1975, the procedure for consulting regional and local bodies has become more "transparent" insofar as the latter are informed of the site proposal(s) through the various regional, département and local bodies, and progress in the discussion may depend on concurrent developments in the preliminary studies *.

In this context, consultation of regional and local bodies has become more important within the overall siting procedure, which is thys, to a certain extent, tending towards decentralisation **.

Minister of the Quality of Life. The list includes a brief reference statement, a description of the studies carried out and an initial assessment of the events linked with siting of the power plant."

* It should be pointed out, however, that the sites proposed after the 1975 Consultation was set up are no longer submitted all at once but from now on are notified to the regions one by one, case by case.

** The intended move towards decentralisation is clearly apparent in the creation of the Regional Consultation procedure. The public authorities which advocated regional consultation were two departments working on the basis of decentralisation of administrative action : DATAR and the Ministry of the Environment. Since its creation, DATAR, which is a "special duty" administration, has had to play a role of promotion and co-ordination between central and local administrative bodies and to

Once the decisive sequence of operations described above is completed, the DUP application procedure begins, culminating in the public authorities' approval by decree of the final site chosen.

4. Application for a "Déclaration d'Utilité Publique" (Statement that the project is in the public interest)

As a result of several changes in the laws and regulations, the DUP application procedure for conventional thermal or nuclear power plants has been amended. The Minister for Industry sends the DUP application file to the Regional Prefect concerned for examination.

The file contains an explanatory note which acts as the "main piece of evidence" *. This contains a comprehensive justification of the site and emphasizes:

1) "the purpose of the operation envisaged";

2) "the reasons why the project presented has been chosen from among other possibilities. In this respect it is necessary to state explicitly the main provisions of the other projects, which may have been drawn up outside the civil service, for example by associations";

3) "the advantages of the project thought to be determining in spite of possible disadvantages". As the enquiry is held with a view to a DUP, "it must be demonstrated that the proposed operation meets the concept of 'public interest' as defined by the Council of State in its judgements of 28th May, 1971 (Case 'Ville Nouvelle Lille-Est') and 20th October, 1972 (Case 'Société Civile Sainte Marie de l'Assomption')". (This will be described in greater detail when dealing with the developments in the procedure.)

The file is completed by an additional document ** on :

---

assert its role as privileged intermediary between the central government and regional bodies. The Ministry of the Environment, which is in many respects a product of DATAR, "naturally" tends to regard the implementation of environmental policy in terms of decentralisation of the measures taken.

Robert Poujade, who was the first Minister for the Environment, held very clear views on the environmental policy commitment of local authorities and a need for "systematic and persistent efforts towards decentralisation and deconcentration" (see his book entitled "Le Ministère de l'Impossible", published by Calmann Levy, 1975).

* Directive of the Prime Minister of 14th May, 1976.
** Circular of the Minister for Industry, dated 24th August, 1976.

- the architectural aspect of the planned in-
  stallation;

- an environmental impact assessment, already
  mentioned above in the second part of the
  paper;

- the main provisions on nuclear safety and radiation protection (for nuclear power plants
  only).

The file is circulated as soon as it has been submitted to the Regional Prefect, but at this stage it
is aimed at "the elected representatives, persons and
bodies concerned by the proposed operation (parliamentary representatives, regional advisers, members of
economic and social regional committees, general coun-
sellors, mayors, chambers of commerce and industry,
agricultural chambers, trade associations, etc.)".

The file becomes public later on only, when the
public enquiry is held, but in the meantime "maximum
publicity will be given to the project by any appropriate means (press releases, audio-visual means,
distribution of leaflets)".

At the same time, the file will be sent to the
"Service Interdépartemental de l'Industrie et des
Mines" (Industry and Mining Inter-departmental Service) * which will launch "inter-departmental conferences". Depending on the case, these are attended by
10 to 20 departments and last six weeks. They are
aimed at ensuring that the siting project is satisfac-
torily taken into account by each department consulted
within its own sphere (compliance with laws and regulations, various requirements etc.). These are not to
be confused with the inter-administrative meetings at
central government level described above.

The EDF replies to the comments made by the civil
service departments. "Any substantial change in the
file must also be notified to the Prefect or Prefects
concerned for information, especially for the elected
representatives and aforesaid bodies" **.

At the end of the administrative conferences, the
"Service inter-départemental de l'Industrie" sends all
the results and its proposals for the enquiry to the
central government, stating the reasons for the proposals. The Minister takes a decision on the enquiry
after receiving the proposals.

---

* This is the new name given on 12th May, 1976 to the
seven outside departments of the Ministry of Industry, which now
replace the former thirteen mining districts and five electrical
power areas.
   ** Circular already quoted.

It should be pointed out that if it was decided :
to postpone the enquiry * it would very probably be due
to the obligation to take account of appreciable developments in the regional acceptability of the site
rather than any difficulties arising during the inter-
departmental conferences **.

The public enquiry lasts from six weeks to two
months. It consists in placing several registers at
the disposal of the public at prefecture and sub-pre-
fecture headquarters and town halls surrounding the
site. Comments on the project, whose DUP application
file is made public at this stage, are entered in the
registers.

The enquiry also involves setting up an enquiry
commission which "is responsible for collecting as
many opinions as possible and for analysing the arguments put forward by the public in order to be able
to assess the project... and it is responsible for
appraising whether the project is in the public interest in the light of the public interest concept
defined by the Council of State" *** (see changes in
the procedure below).

Once the enquiry is completed, the Commission must
submit detailed conclusions to the Prefect within one
month. EDF then replies within three months to the
comments made during the enquiry and, after this period, the "Service Interdépartemental de l'Industrie"
draws up an overall report which is then submitted to
the Minister. The detailed conclusions of the Commission are then published, but not the overall report
submitted to the Minister.

At the end of this stage, which under the Circular's provisions should not last more than one year,
"the Minister responsible for electrical power must
obtain the final opinions of the various ministerial
departments concerned and in particular that of the
Quality of Life, with a view to submitting to the
Council of State a draft decree stating whether the
project is in the public interest".

The whole siting procedure is completed when the
decree is signed.

---

* This happened in the case of the Port La Nouvelle site,
before the Regional Consultation procedure was applied, and
there has been no progress since that date.
** As already mentioned in the first part of the paper,
the purpose of setting up an inter-ministerial consultation procedure ahead of the inter-departmental conferences was to prevent
the emergence of any major administrative difficulties at the
end of the procedure.
*** Directive of the Prime Minister, already quoted.

The present form of the DUP application procedure, which is probably still fluid, is the result of changes affecting both the concept of public interest, that of public inquiry and more generally, relations between the government and the public.

It should be borne in mind that this procedure forms part of the legislation on compulsory purchase of private property, which was laid down at the time of the French Revolution in 1789. It is based on the principle of providing the means for the public authorities to sanction the fact that their projects are in "the public interest" by law, a prerequisite empowering them to undertake the required compulsory purchases once the DUP is made.

Consequently, the procedure is not compulsory; it is only required if compulsory purchases are to be made. A brief summary of recent changes will show that the procedure now also tends to be used in situations not requiring compulsory purchase and that it is thus progressively acquiring a new purpose and gradually becoming a "procedure prior to the decision-making stage" *, **.

Until May 1976, the "public interest" enquiry had been carried out in accordance with a Decree issued in June 1959. In the case of thermal power plants it usually lasted 30 days. It consisted in making registers available to the public under similar conditions to those prevailing nowadays, but with much less publicity regarding the period during which the enquiry was opened, and both the file and the way in which the Enquiry Commission members were recruited reflected the somewhat formal nature of the enquiry.

Thus, in 1974, an inter-ministerial working group *** prepared a reform on public enquiries ****. The group studied the purpose of public enquiries in order to consider whether they should continue to be inserted in the procedure for drawing up engineering projects. In the light of the progressive changes in the purpose of public enquiries, the group suggested that they should no longer be included in the compulsory purchase procedure but that for a fixed period

---

* Under the formula adopted in the "Sialelli Report" to be described later.
** For a long time now, the DUP procedure had been systematically used by the administration for the siting of thermal power plants. Nevertheless, it is significant that the circular of 24th August, 1976 officially sanctions this situation.
*** Chaired by a General Inspector of Equipment, Mrs. Sialelli, with two sub-groups chaired by Mr. Gascoin, Ministry of Equipment and Mr. Rebière, Ministry of the Quality of Life.
**** In other words not only public interest enquiries but also "de commodo et incommodo" and "hydraulic" enquiries.

of time, they should be relatively independent of all other regulations *. In addition to this proposal, the group drew up a number of recommendations, many of which were incorporated in the official text published in May 1976 **.

As already noted in the section on the current DUP procedure, these texts describe most of the detailed aspects. Although the proposal for rendering the public enquiry autonomous (in other words the creation of public hearings) has not yet been adopted, this is currently being examined by the Government ***.

Another related trend concerns the concept of "public interest", which in fact has undergone a far-reaching change of "denaturation" **** which started at the beginning of the 1970s in the Council of State

---

* The grounds for this suggestion are worth quoting in full :

This "fundamental" reform "ought to constitute the ultimate aim owing to its numerous advantages" :
- from the psychological point of view, with respect to public opinion, because it would draw attention to the enquiry by up-grading it to a kind of common-law procedure prior to decision making for numerous categories of administrative acts. It would thus have the merit of contributing to spread the more general concept of a "procedure prior to the decision" in French public law;
- from the doctrinal and legal points of view, in that the enquiry would no longer centre around the guarantee of individual property only, but would appear as an instrument available to any citizen in order to uphold his right to enjoy collective assets (natural assets, quality of urban and rural landscapes etc.) and for environmental conservation. It would thus contribute to bring the French Government's attitudes and actions vis-à-vis the population closer to the more effective ones of the Anglo-Saxon countries and Switzerland;
- from the administrative point of view, by standardizing to a certain extent the various types of enquiries, simplifying the procedure for those who have to use it and encouraging the latter to seek the most suitable ways and means of promoting, through appropriate channels, the supply of information to the public and the latter's participation in preparation of projects "upstream" of the procedure, in other words, as soon as the studies are started.
** See Annex II.
*** The Minister for the Quality of Life has entrusted this task to a Committee chaired by a Counsellor of State. The Committee must submit "a report organising the public hearing procedure, on the basis of which a regulatory text will be issued..." (information given by the Minister during the Parliamentary debate on the Nature Protection Act,J.O. N° 24 AN, 23rd April, 1976).
**** According to André Homont, "l'Expropriation pour cause d'utilité publique" Librairies techniques, 1975.

case law. In view of the scale and increasing complexity of the projects submitted for DUP, the Council of State has been obliged to specify the criteria for assessing "public interest" and to take a decision with reference to the economic and social implications for each project *.

Thus the public interest enquiry seems to serve as one of the means of drawing up an assessment of the social and economic implications of each project. This approach emphasizes the future role which nature protection associations might have to play in liaison with the environmental impact assessment, with a view to integrating environmental aspects into the general appraisal of the consequences **.

In some cases, impact assessments may be supervised by the Ministry of the Quality of Life through its "central environmental work-group" and regional work-groups ***.

In this field, associations are liable to intervene actively in future, particularly in the preparation of alternatives to the impact assessments submitted by the main contractor and, more generally, they could act as a driving and leading force during local discussions on the project's repercussions ****.

---

* Since then, the Council of State has unhesitatingly applied the "cost-benefit" method, "which has become the golden rule in court proceedings when dealing with compulsory purchase" (Case Mr. & Mrs. Ellia, 24th January, 1975, case law record).
** This is only conjectural insofar as very little experience has been acquired on public interest enquiries including an impact assessment, and also because the decree implementing Article 2 has not yet been signed.
*** The procedure for consulting the central or regional work group will probably not be systematic. This will be specified in the decree implementing Article 2. In the meantime, it should be noted that in 1975, the Minister for the Quality of Life estimated that the number of projects likely to be subject to an impact assessment would range between 500 and 1,000 per year. Obviously, in view of the means at their disposal, these workshops cannot examine all these projects, and even less undertake a counter-valuation (for further details see Nungesser report N° 1764 and the minutes of the Senate sitting on 18th May, J.O. N° 285).
**** Along these lines, the Secretary of State for the Environment wrote in "Le Monde" on 29th July, 1976 : "The most important factor in the future of environmental associations must be, however, the substantially formal nature of impact assessments. These associations will thus acquire a very broad field of action : they will be able to appraise impact assessments, to circulate them widely and to organise debates on the subject...".

Appendix I

COMPOSITION AND FUNCTIONS OF REGIONAL COUNCILS *

These Councils were set up under the Act of 5th
July, 1972 on the creation and organisation of regions.
The bodies involved are the Regional Council and the
Economic and Social Committee.

(a) Pursuant to Article 5 of the Act, the Regional
Council is composed of :

i) the elected deputies and senators of the
region;

ii) representatives of the local community elected
by the general councils;

iii) representatives of the built-up areas in each
département of the region.

The seats must be allocated in such a way that
the number of parliamentary representatives does not
exceed that of the representatives of local communities
and built-up areas.

Functions of the Regional Council:

Of the two bodies, it is the only one with powers
of decision. Nevertheless, its powers are often not of
a "decisive" nature **, even if it intervenes as a
civil service body.

The Regional Council's discussions are decisive
only in certain fields where, in most cases, it shares
its powers with other bodies under "conventions",
"agreements", "participation arrangements", etc. The
only field where the decisions are truly unilateral
but remain within the framework of the law is that of
budget and taxation.

Where the Committee intervenes as a body representing the economic consultative government department, the Council's discussions are then always consultative.

---

* It seems useful to outline the functions of these Councils in order to dispel the ambiguities raised by the concept
of "regional institution" in France. The information has been
taken from "les Aspects Administratifs de la Régionalisation",
Institut Français des Sciences Administratives, Cahier n° 10,
Editions Cujas, 1974.

** A distinction must be made from the legal content point
of view between discussions with a "decisive" nature and non-
decisive discussions (expressing wishes), but both are "discussions" (according to F.P. Benoit, Collectivités locales, Dalloz
1973, quoted in the reference work).

(b)   Pursuant to Article 13 of the July 1972 Act, the Economic and Social Committee consists of representatives of the economic, social, trade, family, educational, scientific, cultural and sport bodies and activities of the region. As it represents the various occupational categories, the Committee has only consultative functions.

The Act also specifies that these consultative powers are always applied prior to the decisions made or opinions issued by the Regional Council, which alone may commit the region. Thus there is no competition between the two bodies.

Appendix II

OFFICIAL PROVISIONS DIRECTLY OR INDIRECTLY LINKED
WITH THE PROCEDURE FOR SITING THERMAL POWER PLANTS

The aim is not to give a full description of the
provisions but to list them and to quote certain extracts referred to in the main body of the paper.

- Decree N° 76 432 of 14th May, 1976 amending
  Decree N° 59 701 of 6th June, 1959 : this
  Decree was published in the Official Journal
  of 19th May, 1976 and mainly concerns the
  enquiry procedure prior to the DUP.

- Directive of 14th May, 1976 on the supply of
  information to the public and the organisation
  of public enquiries.

- Act of 10th July, 1976 on the protection of
  nature.

Article 2 : the main extracts are as follows :

"Studies carried out prior to the construction
of facilities or works which, owing to their size
or effect on the natural environment, may harm
the latter, must include an environmental impact
assessment of the consequences."

A decree of the Council of State specifies the
rules for implementing this Article. In particular,
it states that :

"... the content of the impact assessment, which
shall include at least an analysis of the initial situation of the site and its environment, a
study of the changes likely to be caused by the
project and the proposed steps for removing, reducing and, if possible, compensating the harmful consequences for the environment; ..."

As already mentioned, for nuclear power stations,
the impact assessment is summarised in an outline of
the chief aspects of the project (see the third section
of this paper).

"... it shall also fix the conditions under which
the Minister for the Environment may request or
receive any impact assessment for the purpose
of giving an opinion..."

Article 40 :

"Duly registered associations which have carried
out statutory activities relating to the protection of nature and the environment for at least

three years may be officially approved by the
Minister responsible for the protection of nature
and the environment."

Any association aimed at the protection of
nature and the environment "may start proceedings
before the administrative courts for any allega-
tion relating thereto..."

Circular of 24th August, 1976 on the adjustment
of the examination procedure prior to the DUP for conventional thermal or nuclear power plants. It is
addressed by the Minister for Industry to the Prefects
and Heads of Interdepartmental Services for Industry
and Mining. It was published in the Journal Officiel
of 24th September, 1976.

CHAPTER IV

DEVELOPMENTS IN SITING PROCEDURES AND POLICIES
IN THE FEDERAL REPUBLIC OF GERMANY

1. The Federal Republic and its Legal System

    The FRG comprises eleven Länder (regional govern-
ments) of varying size and population density. Table 1
presents their area, population and population density.
Putting aside the city Länder of Bremen, Berlin and
Hamburg, it may be noted that the population densities
of North Rhine - Westphalia (503) and Saarland (427)
are the highest in Europe, exceeding that of the
Netherlands, the most densely populated European
country (367).

Table 1. Area, population and population density
          of FRG Länder

LAND	AREA (Km^2)	POPULATION (x 1000)	AVERAGE POPULATION DENSITY
Baden-Württemberg	35,751	9,152	256
Bavaria	70,546	10,810	153
Bremen	403	716	1,775
Berlin	480	1,984	4,135
Hamburg	753	1,717	2,280
Hesse	21,112	5,549	263
Lower Saxony	47,429	7,238	153
North Rhine-Westphalia	34,056	17,129	503
Rhineland Palatinate	19,837	3,665	185
Saarland	2,569	1,096	427
Schleswig-Holstein	15,678	2,582	165

    Laws enacted in the Federal Republic are either
Federal or Land laws. In principle, Federal laws are
implemented by the Länder as if they were Land laws.
Specific Federal laws are, however, enforced on behalf
of the Federation. With the approval of the Länder
parliaments, the Federal Government may enact general
administrative procedures to ensure the uniform implementation of Federal laws in the Länder. In special

99

cases - e.g. in the case of the Federal Atomic Energy
Law - Federal supervision covers not only legality,
but also enforcement.

The citizen whose rights have been violated by
the public authorities has the constitutional right to
have recourse to the courts. With respect to administrative acts of the Federal, Land or local authorities
the citizen has recourse to the administrative courts
where he can seek the annulment or the enforcement of
an administrative act. Action for annulment may be
brought not only by persons who were directly affected
but also by third parties whose rights have been indirectly affected. Thus neighbours to a site which has
been acquired for building a nuclear power plant may
initiate proceedings against the developer, if their
rights have been violated by the granting of the licence.

The structure of a Land is as follows : the Land
is divided into a few Districts which include a number
of city and county authorities, the latter of which
in turn are divided into communities. Thus the Land
of North Rhine Westphalia (population about 17 million)
has three Districts. A District has up to 14 county and
city authorities. A county has up to 25 communities.

With respect to land use planning, District authorities are given by the Länder the responsibility
to carry out some of their functions, and municipalities are all important as they are responsible under
the Federal Town Planning Law for the binding land
use plans (see below).

2.   Planning Legislation in the FRG

    a)   The Federal Law concerning Regional Planning
          (1965)

This law was enacted as a framework regulation
setting the principles of regional planning for the
whole of the Federal Republic. The Länder may put forward other principles but they must not be in conflict
with those of the Federal Law. The primary aim of the
latter is to guide the development of structural patterns in the Federal Republic in such a way as to
ensure the free development of the individual within
the community while allowing for the economic, social
and cultural development of the country. It is concerned, for example, with the living conditions in
cities, the maintenance of agricultural land and for-
ests, the income of rural populations, the development of substandard areas, and the control of water
pollution.

These principles were given a tangible form in the Federal regional planning programme which is intended to contribute to the development of the regions while maintaining a healthy environment, and to the reduction of disparities in living conditions among regions. The programme is considered as a blueprint for action. Its effectiveness depends on the extent that its goals are adopted by the Land and local governments which are not obliged to do so because the Programme is not by itself a Federal Law.

b) Planning Laws of the Länder

Even before the enactment of the Federal Law, planning laws were in force in most of the Länder. Although the Federal Law serves as a uniform outline for the whole of the Federal Republic, planning legislation in the Länder varies considerably regarding principles, legal apparatus and organisation.

Most of the Land planning laws refer to the siting of major energy facilities and the protection of the environment. In most cases the emphasis is on security of electricity supply at a reasonable price. There is however a trend towards increasing protection of the environment.

The Länder use in common or in general two instruments for translating the content of the Federal and the Land planning laws into practice : Programmes and Plans. The former are usually an outline of policies, principles and goals, while the latter are enacted on the basis of the former and describe the land-use requirements of area and land planning. Not all Länder use both instruments.

The Federal Law concerning Regional Planning obliges the Länder to set up the legal basis for regional planning. The Länder, recognising that land use plans may become more explicit if they cover smaller areas, have provided, through the Land laws, that plans for smaller areas are prepared by District authorities (district development plans).

Most land planning laws allow for the participation, during the preparation of programmes and plans, of advisory boards formed by representatives of political parties, of Chambers of Commerce and industry, of trade unions, of communities, associations, of planning experts, etc. None of the Land planning laws makes provision for direct public participation.

Under the umbrella of the Federal Law on Regional Planning the planning goals of the Länder are taken into account by Federal and Land authorities, local authorities and planning officials. The plans are not legally binding and hence the private citizen cannot

101

challenge them to the administrative courts for judicial review. The Länder have repeatedly stressed in
their Land Planning laws the binding effect of their
programmes and plans on the district and local authorities. It follows, therefore, that planning officials
responsible for the implementation of the programmes
and plans should abide by the prerogatives of the planning guidelines. This, however, had not been explicit
and hence there was a noticeable lack of obligation
on the part of subordinate planning officials to adapt
their plans to previous higher level planning.

c) The Federal Law for Town Planning

Community area planning under the Federal Law for
Town Planning provides the final and definite shape
to the plans that have been prepared by the Land and
the Districts. Neither type of plan, Land nor District,
is by itself a law which binds or benefits the private
citizen. Public law obligations arise for the individual citizen in connection with land and property use
only as a result of community area planning carried
out by the local authority in accordance with the
Federal Law for Town Planning. This law gives considerable freedom and responsibility to local planning
authorities when they prepare their plans, although it
states that these plans should not contradict the Land
and District plans.

Under this law local authorities prepare :

i) The Preliminary Land Use Plan, which is a declaration of intent on the part of the local authority
as to the use of the community land. This plan is not
binding to the public and is neither a legal rule or
an administrative act.

ii) The Binding Land Use Plan, which is based on
the previous one and which is binding for the citizens
and property owners to a very considerable extent. It
deals with the zoning of the land of the community,
the area which is allowed to be built within a property, the provision of transport, of recreational
areas of community infrastructure, etc.

According to the Federal Law on Town Planning,
local authorities when developing their land use plans
should take into account a large number of specific
considerations which serve as guiding principles for
ensuring an orderly urban development, a socially
just use of the land for the benefit of the whole of
the community, and a healthy environment. The law es-
tablishes that when community land-use plans are being
drawn, public and private interests should be taken
into account and balanced against each other. This
leads to the participation of a large number of officials and of representatives of various interests.

The participation of the public has been strengthened
by the latest (1976) version of the text of the Law.
The public can participate in public hearings and in
public disclosure proceedings.

Local authorities are given the responsibility of
organising public hearings. The law stipulates that
not only land-use plans but also planning goals and
targets should be scrutinized by the public during the
hearings.

The disclosure proceedings are confined to the
community land-use plans. The draft plans are displayed
publicly for one month. During that period the public
may submit its reservations and suggestions. If there
are objections, then local authorities will inform the
intervenors of their findings. If the reservations of
the public have not been dealt with by the local authority, they should be transmitted to the higher planning authority (District). The latter examines them
from the legal point of view. If there is no problem
with the letter of the law or with procedures, the
public's reservations may not be taken into account.
Thus the public can dispute such decisions before the
administrative courts only after they have been included in the final community land use plans.

In principle the construction of a building is
permitted only if the site has been specifically identified for that purpose in the local final land-use
plan.

According to the law there are three possibilities
for obtaining a construction licence :

i) The site is identified for this purpose in the
final local land-use plan.

ii) The potential site lies in an area for which
there is no final land-use plan, but which is developed
in such a way that the permit can be issued unless the
nature of the immediate neighbourhood does not allow
it.

iii) The potential site lies in an area which is
not covered by a final land-use plan and which is not
developed. In that case a licence may be issued only
when advantage can be taken of the special exemptions
allowed by the law.

3.   The Siting of Major Energy Facilities in the FRG

    a) With respect to Länder planning legislation

    Given the increasing difficulty for finding sites for
major energy facilities acceptable by the local population,
and the Länder governments' concern for meeting the demand of

energy, Länder planning authorities have laid considerable emphasis on solving the siting problems. Some
of these attempts are described briefly below.

i) Baden Württemberg

The Land has enacted a "Special Development Plan for Energy Facility Sites". This special
Plan sets aside 14 sites for power stations and pro-
hibits their use in the meantime for any purpose which
could prejudice the construction of power stations
when the need for them arises. The Plan was drawn up
by the Land authorities with the assistance of the
Land Planning Board, the communities involved, the
neighbouring Länder (Rhineland Palatinate, Hesse and
Bavaria), the Federal Government, and the neighbouring
countries (Switzerland and France).

The 14 sites were chosen among a list of 41. Each
of them was studied by an interministerial working
group which ascertained their suitability with respect
to water supply, reactor safety, protection from emissions and radiation.

The 14 sites are plotted accurately on the official 1:25,000 map, and their area determined. Twelve
sites are set aside for nuclear power stations. The
Plan ensures about three times as many sites as
will probably be needed by 1985. It also includes
possible alternatives as it is not considered possible
that licences can be obtained for all 14 sites.

This is the first formal plan for securing sites
for major energy facilities in the Federal Republic.

Securing a site, even in a legally binding special
Development Plan, does not mean that a power station can
be definitely built there. The final decision is taken
within the established licensing procedure.

ii) Bavaria

The Bavarian Land Planning Law provides for a
Land Development Programme for the entire area of the
Land. Energy supply is explicitly considered in the
Programme giving the emphasis to meeting the probable
demand in electricity and oil refinery products.

iii) Lower Saxony

A Land Development Programme for 1985 was published in 1973; the Programme contains the following
statement with respect to major energy facilities.

"... The identification of sites for future conventional and nuclear power stations is becoming increasingly difficult. The conflict between the need
for more electricity-generating stations and for protection of the environment can only be resolved through
co-operation of those involved. Since 1972 a working

104

group of representatives from several ministries have been examining in close co-operation with the Federal, Land and local authorities and representatives of the energy supply industries the suitability of various sites. Of the 16 sites examined, preliminary results have been obtained for Emden, Cuxhaven and Grohnde. Once the examinations are completed, suitable sites for power stations will be indicated in the Land Development Programme and in the District Development Programme.

iv) <u>North Rhine Westphalia</u>

Developments in North Rhine Westphalia are described to considerable extent in the following sections.

In general way, the siting is included within the framework of Land planning varies among Länder. Baden-Württemberg enacted a "Special Development Plan for Energy Facility Sites" in 1976, whereas in Bavaria and North Rhine Westphalia, it is expected that during 1977 plans will become binding.

b) <u>With respect to the Federal Town Planning Law</u>

Energy facilities may be sited only in areas identified by the Federal Building Law as "industrial" or "special". Furthermore, not all industrial areas may accommodate them. The decision is based on the legal provisions of the Federal Immissions Control Law and, for nuclear power installations, on the Federal Atomic Energy Law. The absence of land-use planning objections is only one of the preconditions for the site.

Major energy facilities can be built in developed areas only when they replace similar facilities at the same site and are of comparable size. In other than industrial areas, each case is decide specially basis. The trend in legislation is, however, towards reducing the possibilities of siting in areas other than those indicated as suitable for that purpose in the plan.

A further indication of the importance of local land-use plans is that they provide the only instrument for early public participation in the licensing procedure. If an application for building in a area not identified as suitable in the local plans is put forward, the public will have the possibility to participate only at the very late stages of the licensing procedure. The 1976 revision of the Federal Law for Town Planning stipulates that the public should be involved in hearings at an early stage of the preparation of the local land-use plans.

There are very few environmental protection measures which are within the responsibility of local authorities of the communities. Radiation protection air pollution and noise abatement are regulated by Federal laws enforced by the Länder. There is, therefore, the probability that the lack of environmental expertise of the local authorities will lead to local binding land-use plans which have been prepared without due care to environmental protection.

To make up for this lack of expertise, the Land of North Rhine-Westphalia obliges the Land Factory Inspectorates, which are responsible for enforcing the pollution control and anti-noise legislation, to take part in the local land-use planning procedures as representatives of the public interest. Their main responsibility, as determined by the Land Minister of Social Affairs, is to examine the compatibility of the community land-use plans submitted by the planning authorities with the principles of the Federal Immission Control Law. The examination is centered around the location of industrial and commercial areas with respect to residential areas, and the distances of particular industrial and commercial facilities from residential areas. For example, the Ministry's directive specifies that coking plants and petrochemical installations should be at least 1,500 metres from residential areas, oil refineries at least 1,200 metres, and conventional power stations of more than 220 MWe at least 1000 metres.

c) <u>Licensing procedures for major energy facilities</u>

A licence is required for constructing and operating such plants. The licensing procedure for nuclear power stations is defined by the Federal Atomic Energy Law, and that for conventional power stations and other major energy facilities by the Federal Immissions Control Law. In both cases, one of the licensing requirements is that the site conform with the prerogatives of the Federal Town Planning Law.

The Federal Atomic Energy Law is applied and enforced by the Länder on behalf of the Federal Government. The law stipulates that the land licensing authority for nuclear energy facilities should be at the highest level (Ministry). Federal Land and community statutory authorities affected by the application take part in the licensing procedure. The procedure followed is that of the Federal Immissions Control Law.

A large number of physical, environmental, social, economic and technical aspects are examined to considerable detail. Radiation protection is examined according to a special Federal Directive.

The Federal Minister of the Interior, who is responsible for the licences of nuclear energy facilities, receives the application from the land licensing authority, consults with other Federal Ministers and the Reactor Safety Commission. The safety recommendations received from these quarters are passed on by the Minister to the licensing authority as directives.

In the course of the examination, the project is advertised to the public and supporting documents are made available to all interested parties. Sometime after the end of the disclosure period, the licensing authority holds a hearing to give objectors and supporters an opportunity to clarify and develop publicly the comments they have submitted in writing.

The licensing authority reaches its decision after assessing the comments of the statutory authorities, officials, the specialists, reports, as well as the directives of the Federal Minister of the Interior and the outcome of the public hearing. The procedure takes one to two years from application to decision. If the facility is approved basically, some 8-12 secondary (partial) licenses are needed dealing with the site, safety measures, installations, machinery, start-up operations, etc.

Decisions taken under the Federal Atomic Energy Law are subject to review by the administrative courts, as any other administrative act.

The Federal Immissions Control Law is applied and enforced by the Länder on behalf of the Federal Government. Licensing authorities for major energy facilities vary among Länder and for the different types of facilities. Usually they are not at Ministerial level, as is the case with nuclear power installations.

According to the Federal Immissions Control Law, a licence to construct and operate a major energy facility may be granted only if it has been shown that

- harmful environmental impacts for the public in general and the local community in particular will not arise;
- measures for the control of harmful environmental emmissions will be taken, using the best available technology;
- waste produced by the facility will be recycled and if this is technically or economically impossible, waste will be disposed of in accordance with existing regulations;
- other public laws or work protection rules do not stand in the way of the construction and operation of the facility.

The law does not elaborate on the above general concepts used to describe the licensing criteria. It leaves the specific provisions required for the implementation of the law to the Federal Government which provides the general administrative procedures and standards, after consultations with scientific representatives, the industry involved, the transport sector, the affected citizens and the Land authorities. In general, the Law deals with emission standards which should not be exceeded and levels of immissions which should be kept, given the state of pollution control technology. There are two Administrative Directives which affect the siting of major facilities:

 i) technical instructions for maintaining air purity
 ii) technical instruction on anti-noise measures.

Once the licensing criteria have been met, the applicant has a legal claim to obtain a licence. This however does not include water use. If, for example, river or underground water is needed for cooling, a special licence should be obtained. Another special licence is needed for waste disposal (ashes, sludge, etc.).

The licensing authority, according to the Federal Immissions Control Law, advertises the proposed project in the Official Gazette and the local press. The application and supporting documentation is made available to the public for a period of two months. During this period, objections to the project may be lodged with the licensing authority. Following the disclosure period, the licensing authority holds a hearing at which it discusses the objections with the applicant and the objectors.

The licensing authority reaches a decision on the basis of the examination it has carried out, the opinions of statutory authorities, and the results of the hearing. It is then published and communicated to the applicant and the objectors, unless they are more than 300, in which case the decision is advertised to the public for two weeks.

4.   Implementation of the Federal Government's Energy Programme

a)   The programme

Shortly before the oil price crisis of 1973 the Federal Government submitted its long-term energy programme. This had to be amended in the light of the recent energy and economic developments. The programme proposed in November 1974 projects a lower rate of energy consumption (about 9% less for 1985 than what was expected in the 1973 programme) for the Federal Republic. Table 2 indicates the breakdown of primary energy use in the Federal Republic, as forecast in the 1974 Energy Programme.

Table 2. Primary Energy Use for the FRG

Energy Resource	1973		1980		1985	
	Million tons bit.coal.eqt.	%	Million tons bit.coal.eqt.	%	Million tons bit.coal.eqt.	%
Petroleum	209	55	221	47	245	44
Bituminous coal	84.2	22	82	17	79	14
Natural gas	38.6	10	87	18	101	18
Lignite	33,1	9	35	7	38	7
Nuclear Energy	4	1	40	9	81	15
Other	9.7	3	10	2	11	2
Total	378.6	100	475	100	555	100

1 Million ton bituminous coal equivalent = 7×10^{12} Kcal

The main characteristics of the Energy Programme
are the following :

i) Oil, despite efforts to reduce dependence on
it, will remain the most important energy resource.

ii) Natural gas will be used to reduce dependence
on oil in the short term; along with natural gas piped
from the Netherlands and the USSR, large deliveries
are expected from the Ekofisk and other North Sea
fields, Algeria and Iran.

iii) Nuclear energy will be used to decrease dependence on oil in the longer term; it is expected that
electricity production from nuclear energy will be
increased to more than 20% of the total in 1980 and
to 30% in 1985. This will require a nuclear power plant
capacity of 13,500 MWe by 1980 and of 24,000 MWe by
1985.

iv) The use of bituminous coal as an energy source
will be stabilised to current levels.

This goal can be reached only with government
support, particularly regarding the use of coal for
electricity generation. For this purpose the Federal
Government enacted the Third Electricity Generation
Act in 1975 and amended it early in 1976. The Federal
coal policy is however hampered by the difficult com-
petitive position of German bituminous coal resulting
from its unfavourable geologic structure and the high
wages that must be paid for its mining.

With regard to the protection of the environment,
the first amendment to the Energy Programme indicates
that energy and environmental policies must not be

conflicting. It does not however go beyond general statements on energy supply and the protection of the environment. The need for energy facilities sites is mentioned as an important factor and doubts are expressed on whether the necessary sites for implementing the Energy Programme could be secured in time following the licencing procedures in force, particularly for nuclear and coal-fired power stations.

b) Refineries

The forecast for oil consumption is that it will rise by 30 million tons between 1975 and 1990. Oil consumption dropped between 1973 and 1975 from 145 to 125 million tons and amounted in 1977 to 133 million tons. Refineries were operating in 1973 in Germany at about 83% capacity (approx. 145 million tons), whereas in 1977 at about 65% capacity (approx. 155 million tons).

This excess refining capacity is to be found throughout the Common Market countries. In the Federal Republic it is not so much a question of siting new refineries as it is of extensions of petrochemical plants associated with old refineries.

c) Nuclear power stations

The situation of nuclear power stations in the Federal Republic is as follows (1977).

i)	In operation	14 units	total capacity	7,400 MWe
ii)	Under construction (most without final licence)	10 units	total capacity	10,700 MWe
iii)	Licensed but stopped by courts	3 units	total capacity	4,100 MWe
iv)	At licensing stage	5 units	total capacity	6,100 MWe

Most nuclear power stations at the construction stage are still involved in administrative legal suits and modification.

d) Coal-fired power stations

The Federal Energy Programme and amendments brought to it include the development of coal-fired generating capacity of 6,000 MWe. Coal is given an edge over heavy fuel oil, the impact of which is restricted by law.

The structure of the 6,000 MWe coal programme is as follows (1977) :

i) In operation	1 unit	total capacity	650 MWe
ii) Under construction	5 units	total capacity	3,900 MWe
iii) Approved	5 units	total capacity	3,000 MWe

A list of future power stations is still subject to approval. Their construction will depend on the development of the growth of electricity demand.

Many of the problems associated with the 6,000 MWe coal programme can be traced to the fact that as bituminous coal is in general uneconomic when compared to residual oil, it should preferably be burnt in situ. As more than 80% of the coal deposits in the Federal Republic are to be found in the Ruhr Valley, most of the power stations should be built there.

e) Lignite power stations

Lignite fields in the Federal Republic are found mainly in North Rhine-Westphalia (approximately 87% of the output), smaller fields are worked in Lower Saxony (4%), Hesse (3%) and Bavaria (6%). Conversion of lignite into electricity is confined to the areas where it is found; the mines are solely operated on the surface. Lignite power stations are located in North Rhine-Westphalia exclusively in the Rhenish lignite area around Aachen-Cologne-Mönchengladbach. The Federal Energy Programme provides for only a small-scale development of lignite power station capacities. Major difficulties have not yet been encountered in the construction of lignite power stations in the Rhenish lignite area. This is mainly due to the fact that lignite power stations are sited in rural areas where air pollution is considerably less than in the congested industrial areas of the bituminous coal fields in the Ruhr Valley.

f) Planning for energy facility siting -the example of North Rhine-Westphalia

The key words for solving the problem of siting are early identification of sites. In its 1973 energy programme the Federal Government described measures that were designed to ensure the timely identification of suitable sites for the development of all adequate energy supply capacity. In the first amendment to the 1974 energy programme the Federal Government repeats its intention and states that surveys should be carried out for sites for the 1985 and 1990 time horizons. The Federal Government stressed also in its 1974 Federal Planning Report the need for facility sites. The report contains a map of existing power stations and of those planned for 1980.

Furthermore, the 1976 Federal Government Report on the Environment contains the following with respect to the provision of sites for major energy facilities.

"... suitable sites for industrial facilities of more than regional importance, particularly major energy facilities, must be duly identified. In cooperation with the Länder, the Federal Government will work out a scheme for planning and securing sites, including supply and disposal facilities in the energy field, within long-term regional development plans with the participation of the affected population. This work programme comprises general requirements and assessment data for an early examination of possible sites, a survey of land-use plans, agreements on projects near to borders between Länder or with neighbouring countries, and a long-term siting policy within EEC."

The provision laid the cornerstone for the development of a new land-use planning approach which permitted the Land planning authorities to ensure the smooth implementation of major projects of political and economic importance by influencing community land-use planning. Thus the Land government is now able to use the recently acquired legal machinery for siting large industrial projects, including power stations.

The Prime Minister of North Rhine-Westphalia, in his capacity as Land planning authority, prepares the Land development plans which provide the principles and aims of area and land-use planning for the whole land. These principles and aims are established in accordance with the Federal Law for Regional Planning of 1965, but further principles are chosen by taking into account the special regional and structural conditions of North Rhine-Westphalia.

The principles and aims of a Land Development Plan are incorporated in the district development plans which are elaborated by the district planning authorities. District plans are in turn incorporated into community land-use plans. The Federal Building Law makes it obligatory for community land-use plans to conform to the expressed aims of District and Land planning. Consequently, the Land Planning Law requires that whenever communities intend to alter their land use plans they disclose their intentions to the district planning authorities and request to be informed of any existing planning goals which are likely to be affected. The Land Government may, therefore, issue planning instructions to the communities for projects of major economic importance. There are two conditions which must be fulfilled for identifying energy sites in the communal land use plan. First, they must be considered as "developments of particular importance

112

for the economic structure of the Land" and, second, they must be designated in the district development plans.

The VIth Development Plan was recently drawn up. Sites for large industrial projects and for power stations proposed by Land planning bodies, district and local authorities, trade and industry have been examined with respect to a set of criteria. The proposing authorities did not carry out any evaluation of the probable impacts or any examination of whether or not the siting of the proposed facilities conforms to existing legislation.

The main criteria used for incorporating a proposal into the VIth Development Plan were the following :

> Geology and seismology
> Protection of agricultural land
> Protection of forests
> Protection of mines and quarries
> Protection of landscape
> Protection of recreational areas
> Protection of residential areas
> Infrastructure and economics
> Labour market
> Waste disposal facilities
> Mineral wealth
> Water supply and regulation
> Water mains and drains
> Topography
> Immission control

Additional criteria for power plant sites were the following :

> Position regarding coal consumption
> Position regarding electricity grid and peak
>   hour consumption
> Meteorology (diffusion of pollutants, dissip-
>   ation of cooling tower waste heat)

For nuclear power plants further site criteria are given :

> Population density
> Transport routes
> External influences (aircraft accidents,
>   explosions, etc.).

The examination based on the above criteria, enquiring whether the proposed site should be included in the VIth Development Plan, determined only those sites which, in principle, seemed to be suitable for such facilities, particularly from the environment protection point of view. As the examination was a brief one, no decision was taken as to whether particular projects in the proposed areas could be

113

legally admissible. This was left for the subsequent
stages of the planning procedure, namely the district
and communal-land use plans. Considerable effort was,
however, put into evaluating their acceptability under
the Federal Immissions Control Law and the Federal
Atomic Energy Law, which could forbid the issuing
of construction and operation licences and thus
jeopardise the implementation of energy and economic
policies.

With respect to the Immissions Control Law, two
criteria were used at this stage :

  i) the levels of concentration of sulphur oxides,
     particulates and fluorides in the proposed
     area as determined by the Land Agency for
     Immissions Control long-term measurements, and
 ii) the distance between the proposed site and
     urban areas around it.

With respect to the Federal Atomic Energy Law,
the guidelines used for the evaluation of the proposed
sites for nuclear power stations were those of the
"Evaluation Criteria for the assessment of nuclear
power station sites from the standpoint of reactor
safety and radiation protection" worked out jointly
by the Federal Government and the Länder. The sites
were evaluated on the basis of the following criteria :

   i) meteorology;
  ii) hydrology;
 iii) population density;
  iv) transport routes;
   v) external influences.

This was only a preliminary examination which could be
followed at a later date by an exhaustive investigation
in case of an application for a licence.

Competent ministries examine the other site criteria listed above for each proposed site. Those sites
which seem to be suitable for the proposed facilities
are included in the VIth Land Development Plan. A draft
form of the Plan was sent from the Land Planning authority to the district planning authorities and the
appropriate local authorities for comments. After
assessing their opinions, the Land planning authority
will draw up the Final Development Plan, which must be
submitted to the appropriate Land Parliamentary
Committee and the Land Cabinet for approval.

Once the development plan is adopted and published,
the district planning councils will be asked to incorporate the objectives of the plan into district development plans. When this is done, the local authorities
will be asked to draw the necessary land-use plans
which form the legal basis for any claim on land for

development. It is expected that local authorities, who have been invited to participate at an early date in the drawing up of the VIth Development Plan, together with the district planning authorities, will cooperate at that stage and there will be no need for the Land government to issue planning orders to communities to acquire proposed sites.

In this way, it is expected that North Rhine-Westphalia would have by 1980 enough sites to accommodate its needs for major energy facilities. Licensing procedures will then be simplified.

CHAPTER V

SOCIAL AND ECONOMIC IMPACTS ON LOCAL
COMMUNITIES OF THE SITING OF MAJOR ENERGY FACILITIES:
THE CASE OF NUCLEAR POWER PLANTS

1.   Introduction

The number and size of industrial plants has been
growing at an increasing rate for the last 20 years,
and they have often taken on new forms for which former
experience and even common sense were of no value in
forecasting their impact. At the same time a desire
to protect the environment has been growing stronger
and more widespread. In the resulting general climate
it has been necessary to make fundamental changes in
the procedures for authorising industrial facilities,
especially from the standpoint of standards of construction and operation needed to safeguard the quality of life.

For an industrial plant to be accepted even when
the inconvenience which it causes the population goes
beyond the limits laid down by regulations, it is now
necessary for it to be able to meet the following
criteria of public interest :

- nationally :

  . the creation of the facility must meet an
    actual public need (as to capacity and date
    of coming into production);

  . the socio-economic balance must be positive;

  . the siting must be that which ensures the
    best socio-economic balance;

- locally :

  . the socio-economic balance must be positive.

The population has always asked, and more recently insisted, that it be given an opportunity to defend
its legitimate rights by participating in the planning and the regulatory procedures for each project;
this applies especially to the local population exposed to the greatest inconvenience because of its nearness to the plant. The effect of all these conditions

has been to make the procedures for authorising industrial plants considerably more time-consuming.

This situation was taken into consideration in
the studies undertaken in the course of finding out
how ecological and energy needs could be reconciled.
The studies highlighted above all the desirablity,
from the standpoint of hastening the issue of the
licence to establish an industrial plant, of ensuring
that it will be accepted by the population, and especially the local population; and it was felt that the
best means of achieving this was undoubtedly to give
the local population, in the form of collective benefits, the compensation to which it could justly consider it was entitled in return for the collective
inconvenience to which it alone is exposed to the same
extent because of the nearness of the plant; in a
word, it is necessary to improve the local socio-economic balance so that it shall no longer be much less
favourable than the nation-wide balance; incidentally,
legislation allows the conferment of collective benefits.

These general conclusions are especially
relevant in the case of nuclear power plants, for which
acceptance by the population is a problem of the greatest topical importance and one that is particularly
difficult to resolve; the measures already taken in
this instance should, it seems, provide examples of
great instructive value.

2.   Disadvantages of proximity

    a)   Use of Land
The area taken up by a nuclear power plant can
vary within fairly wide limits depending on the number
of units, the method of cooling (open-circuit or atmospheric cooling) and above all local conditions.
For a single unit with a pressurised water reactor
(PWR) of 925 MWe the area taken up may be in the order
of 100 hectares, and for four units of the same type
with cooling towers, 250 to 300 hectares. To these
areas must be added the space occupied by the pylons
of the overhead transmission lines, and that overhung
by the lines themselves; it is difficult to give figures for this as there may be considerable differences
between one site and another.

Where there are land-use plans, a power plant
cannot normally be sited elsewhere than in an industrial area.

For the electricity transmission lines, the ground
between pylons and under the wires can still be used
for cultivation (a common solution in the United States).

At St. Laurent des Eaux in France the plant is on land alongside the River Loire and formerly liable to flooding, and for the first two gas-graphite units, each of 500 MWe, occupied only 40 hectares of cultivable land. This area was sufficiently small in comparison to the total area under cultivation in the commune for the number of farms not to have to be reduced.

At Doel in Belgium the plant occupied some cultivable land but in an area which had formerly been classified as industrial, and for this reason the farmers' protests had no effect.

In the United States the SKAGIT plant (Sedro Wooley, Washington) was built on land (105 hectares) which had formerly been zoned as forest but was later re-classified as an industrial area with the agreement of the local population because of the advantages which the plant represented for local economic development (see 3.B.d. below).

b) Immediate surroundings of the construction site

The construction of a nuclear plant calls for the setting up of a "large construction site" that is large in terms of both the numbers employed and the period of time involved.

In France, the EDF * gives the following figures :

Plant	Equipment	Numbers employed (peak)	Duration (Years)
Fessenheim	1 unit 925 MWe P.W.R.	861	5
Le Bugey	4 units 925 MWe P.W.R.	2600	7
Gravelines	4 units 925 MWe P.W.R.	2600	
Paluel	2 units 1200 MWe P.W.R.	1500	
Creys-Malville	1 unit 1200 MWe breeder reactor	1600	

In France the majority of the construction workers on all the nuclear power plants have hitherto had to be housed near the site (60 per cent of the total); and allowing for families accompanying the workers, the total increase in population is expected to be at the rate of 2,800 for 1,500 workers or 3,000 for 2,000 workers. To this increase in population must be added that caused by the presence of EDF staff to inspect the work and carry out the preliminary training of the staff to operate the plant.

* Electricité de France, a nationalised public industrial and commercial enterprise for the generation, transmission and distribution of electricity throughout French territory.

Arrangements have had to be made to ensure that
such a large increase in population compared with the
local population does not cause serious difficulties
from the point of view of the conditions of life of
the workers and their families, and their accommodation in particular; the admission of their children to
schools; traffic on the roads; the quality of the public, medical, postal, commercial services and so on;
and entertainment and leisure activities, for the lack
of which the workers and the local population in particular might suffer a deterioration in the local social
climate.

These arrangements now come within the framework
of what has been called the "regime for major construction sites".

This regime, which does not only apply to nuclear
power plant sites, is designed to ensure that the
conditions of life of workers on a large site resemble
those of the local population as closely as possible;
the opening of a site must also be the occasion for
adapting the general communal facilities of the surrounding area to the new needs.

These arrangements are made in collaboration with
the local communities concerned and at no cost to the
latter which cannot be covered by the new tax revenue
brought by the plant, i.e. with no increase in local
taxes; this is true even if, in addition to specific
accompanying facilities such as housing and schools, it
is found necessary to provide other facilities in advance to improve the future quality of life, such as
sports fields, swimming pools, etc.

Local communities may be self-financing or may
borrow with the help of the financial facilities given
them, the financing being covered mainly by the large
new tax revenue going to the local communities from
taxation of the plant.

This favourable situation will not be compromised
but rather improved from a regional standpoint, and-
so far as it affects neighbouring communities and land
use planning, by the fact that under the new tax regulations of the Taxe Professionnelle the revenue from
a nuclear power plant is no longer, since 1st January,
1976, reserved for the commune in which the plant is
established but can be used to benefit neighbouring
communes in accordance with the cost to them of their
proximity to the plant.

The numbers of workers employed on nuclear plant
construction sites in Belgium, Switzerland and the
United States are of the same order as in France, as
is the period of time for which a site is open.

In every case hitherto in Belgium and Switzerland, however, and often in the United States, the number of workers to be housed on the site is comparatively small, most of them living in their normal homes, as do their families; they are what are known in Belgium as "semainiers" or weekly workers, and in the United States as "commuters" who are usually willing to work 50 km from home in Belgium and 2 - 2 1/2 hours' journey in the United States.

When these conditions are present the opening of the site no longer results in a large population increase, and the problems of housing and schools, and the corresponding costs, hardly arise. There remain however the important problems of policing and roads resulting from the heavy traffic when workers arrive at and leave the site in their own cars or in buses. In these circumstances one American town, which was receiving large tax revenue solely from a nuclear plant but only incurred small costs in return, was able to decide to reduce its rates of tax *.

The disadvantages of proximity to a plant may naturally extend to communities adjoining that in which the plant is situated, and cause the same difficulties as have arisen, for similar reasons, in operating the plant. The neighbourhood of the site may still suffer the inconveniences of noise, smells, dust, disturbance of watercourses or lowering of water-tables; these last are now, however, limited by the regulations on protection of the environment or by special provisions that may be included in the permit to construct the plant and its ancillary buildings.

c) Vicinity of the plant in operation

The rules now in force ensure the protection of the environment against thermal or radioactive pollution which may result from the operation of a nuclear plant; the provisions are very extensive and very strictly administered. On the essential points of health and safety the inconvenience of a nuclear power plant in operation is thus very limited, and may be even more so as a result of special conditions imposed on the constructor when the construction permit is granted.

The local population is particularly concerned with the following inconveniences of operation : thermal pollution of water, aesthetic disturbance, noise, and occasionally a change in the climate.

Thermal pollution is a source of concern through its possible repercussions on fishing, although those so far experienced have been comparatively slight.

---

* Plymouth (Mass.), Pilgrim plant.

120

The aesthetic disturbances are often the result
of the uncompromisingly modern industrial form and
great size of the buildings of the plant, the increasing number of transmission lines for the electric power
produced, and the difficulty of integrating all these
elements harmoniously in the landscape, especially
where large cooling towers (at least 80 metres high)
are concerned. The designer always tries to resolve
this problem with the help of the most highly qualified
town-planning architects. On a site which was suitable
for such work, one company even went so far as to
lower the level of the plant to make the buildings
less conspicuous: by 20 metres for the plant itself
and 30 metres for the cooling towers *.

Cooling towers may cause noise; important research
aimed at reducing it is underway and can be expected
to solve the problem fairly soon.

Anxiety has been expressed concerning changes in
atmospheric conditions in areas close to large plants
equipped with cooling towers; here, also, important
research has been undertaken on models which has made
it possible to determine the general arrangement to be
followed for cooling towers so as to avoid the effects
on climate that were feared. In any case, it is now
known to be possible to construct cooling towers which
do not produce fog or ice in their immediate neighbourhood in cold weather.

So far as the transport of fuel is concerned, the
annual tonnage needed for nuclear power plants is comparatively small, but such transport does call for
conditions and controls to ensure its safety.

3.   Examples of compensation

   a)   Tax revenue

   i)   Preliminary observations

Here we shall consider only those revenues which
benefit local communities directly and are derived from
taxes paid by the constructor of a nuclear plant, hence
excluding the ordinary local revenue resulting from
economic development brought about indirectly by the
establishment of the plant (which will be dealt with
later) and taxes paid by contractors, their workers,
and EDF personnel.

Nuclear power plants are usually liable to the
same taxes as other industrial establishments as far
as the bases and rates of tax are concerned; but the
tax bases may be reduced for nuclear power plants because of the exceptionally high specific value of their
equipment.

---

* The Leibstadt plant in Switzerland.

The amount of tax paid by a nuclear power plant
in relation to its high value is nonetheless comparatively high; and the amount, usually collected by a
small local authority, considerably increases the budget potential of that authority, which is usually the
only one to benefit from it.

Local communities immediately adjacent to that in
which the plant is situated may suffer disadvantages
from such proximity which at present have no fiscal
counterpart in most countries. This situation has just
been remedied in France now that the "taxe professionnelle" has replaced the old "patente" tax, with the
result that communes adjacent to that in which a plant
is situated may now receive a share of the taxes paid
to the commune in which the plant is situated so that
only part of it remains at its disposal.

ii) The situation in France

In France electric power plants are constructed
and operated by the EDF. This body is required to pay
local taxes on each plant equivalent to those payable
on all other industrial establishments. The taxes used
to include a land tax, and, more important (amounting
to four-fifths of the total), a business activity tax
known as the "patente".

The amounts paid by EDF under the heading of the
patente for nuclear power plants were as follows in
1976 :

Plant	Continuous net output (MWe)	Amount of patente (Frs. million)
1. Chinon	690	8.8
2. St. Laurent des Eaux	975	18.9
3. Le Bugey	540	6.7

which corresponds to Frs. 6,480 per head of population
for St. Laurent des Eaux in 1975.

Under this arrangement the commune in which the
plant was situated enjoyed the tax privilege described
above. The neighbouring communes complained, after the
attempts to settle the matter in a friendly manner
remained without result.

The problem was settled from 1st January, 1976
onwards by the introduction of the "taxe professionelle"
replacing the "patente". Under the new regime there
is a ceiling on the amount of tax revenue which can
go to the commune in which the plant is established,
the balance being paid to a departmental fund for

distribution by the Conseil Général; the communes adjacent to that in which the plant is situated receive some share of the tax, the main criterion for the apportionment being the number of wage-earners at the plant living in each commune. Communes in which a plant had been situated before the change are covered by transitional arrangements so that they will not experience any difficulty owing to changes in their tax revenue in settling any financial commitments they might have contracted earlier.

Quite generally, and not especially in connection with nuclear power plants, the taxe professionnelle has encountered many difficulties in its application, and transitional arrangements have had to be made until a new revised text can be drawn up and placed before parliament. Because of this it has not been possible for us to record the amount of tax revenue received by French communes in 1976 under the new tax; taken as a whole it is likely to be of the same order as in 1975.

iii) The situation in Belgium

In Belgium the two nuclear power plants at Doel on the Scheldt and Tihange on the Meuse were built and operated by a company whose capital is privately owned (by a group of Belgian electricity companies, with EDF participation in the case of Tihange). Each plant is therefore a fiscally independent industrial establishment.

The taxes payable in Belgium by the nuclear power plant consist of :

- tax on the income from the property;
- provincial taxes on motive power and labour;
- communal taxes.

The basis of taxation of the income from the property is calculated in accordance with the value of the installations and equipment, and is in practice approximately proportional to the overall investment (of the order of 5/1000 of that investment). The amount of tax payable is determined by applying to the tax base a rate of what is known as the "précompte immobilier" varying from 40 to 60 per cent according to the site.

The yield of the property income tax is shared between the State, the Province and the commune in approximately the following proportions :

State	30 per cent;
Province	10 to 20 per cent;
Commune	20 to 50 per cent.

The basis for levying the provincial taxes is calculated according to :

- the motive power consumed by the auxiliary plant;

- the number of employees at the plant.

The rates vary in accordance with the budget decisions of the province and the commune.

The amount of the communal taxes can be of the order of B.Frs. 12 million annually for a P.W.R. type nuclear unit of 925 MWe.

On the basis described above, total annual tax revenue from a nuclear unit of 925 MWe may be estimated to have been as follows at the beginning of 1976 :

Province	Property income	B.Frs. 7 to 15 m
	Taxes	B.Frs. 4 m
	Total	B.Frs. 11 to 19 m
Commune	Property income	B.Frs. 15 to 35 m
	Taxes	B.Frs. 12 m
	Total	B.Frs. 27 to 47 m

Discussions are now going on between the tax authorities and Belgian companies concerning the bases and rates of tax to be applied to nuclear power plants, with a view to determining what derogations or relief might be justified by the exceptionally high level of investment required for the construction of this type of plant.

iv) The situation in Switzerland

In Switzerland nuclear power plants are normally constructed jointly by several electricity companies, which form for this purpose a private company owned by them and responsible for the construction and operation. The unit size made necessary by the use of nuclear energy may thus be achieved economically, allowing for the installed capacity of the country and its development.

The company exists for no other purpose than to construct and operate the plant, and it is in accordance with this situation that taxes are levied for the benefit of the local communities.

The company is taxed on capital and income.

The bases and rates of tax are laid down by a cantonal law the provisions of which apply uniformly to all "capitalised" corporate bodies in a Canton; the amount of tax payable is fixed every two years for the current and following years.

Where two or more communes or Cantons are involved,
a part of their territory being occupied by the plant,
the apportionment of the yield of the taxes is determined by a uniform Federal law in the same way as in the
case of tax regulations applying to foreign participation.

For a plant comprising one unit of approximately
900 MWe the annual total of cantonal and communal
taxes is of the order of Sw.Frs.6-7 million, broken
down as follows for the two assumed values of the
share capital of the construction and operating company :

	Share capital	
	Sw.Frs million	Sw.Frs million
Cantonal tax	4.58	5.11
Tax of commune in which situated	1.72	1.47
Tax of all communes in the Canton	0.44	0.49
Total (Sw.Frs. million)	6.34	7.07

The tax revenue of the commune in which the plant
is situated is quite adequate to cover the costs which
the commune may have to bear resulting from the establishment of the plant, the opening of the building
site requiring only comparatively small incidental work
in connection with the accommodation of the workers,
schools and religious establishments.

The tax revenue of the other communes in the
Canton, distributed out of the general tax fund in proportion to the population of each commune, is very
small; this method of apportionment would not suffice
to cover the comparative inconvenience which a commune
immediately adjacent to the plant might have to bear
if it became especially serious; in the case of the
Ingwill plant, planned by the Forces Motrices Bernoi-
ses, studies have already been started to ensure that
a more substantial proportion of the tax revenue will
be available for sharing among all the communes having
a portion of their territory within 5 kilometres of
the plant.

The local communities obviously only receive tax
revenue after the plant has begun operations; during
the construction period, however, they may benefit
from the capital tax which increases in proportion to
the growth of the share capital of the construction
company.

For the Leibstadt plant, for instance, and for a unit of 925 MWe, the taxes paid during the construction work on account of the share capital should gradually rise, as that capital increases :

- from Sw.Frs. 0.8 to 1.6 million per annum for the Canton;

- from Sw.Frs. 0.3 to 0.7 million annually for the commune in which the plant is situated.

v) The situation in the United States

Under the federal system of the United States the tax revenue accruing to local communities as a result of the establishment of a nuclear power plant varies in form and amount from one State to another and even from one place to another within a single State.

The following figures have been compiled from four Environmental Impact Statements :

Clinton plant
Two units 993 MWe, total output 1,866 MWe, recipients of tax : Witt County, Haro township, school district

Amount forecast for 1983 : S4.3 million per annum weighted for inflation

Amount for 1977 (unweighted): $2.5 million per annum.

Vogtle plant
Four units 1,100 MWe, total output 4,400 MWe

Recipients of taxes and annual amounts in 1983 : Schools in the county : $10.53 million
Health scheme : $4.455 million
Total in 1977 :$11.2 million.

Perry plant
Two units 1,205 MWe, total output 2,410 MWe
Recipients of taxes and annual amounts :
School system $34.4 million
Village fund $ 1.1 million
Township $2.1 million
County fund $5.5 million

Jamesport plant
Recipient of taxes : Riverhead school district
Annual amount 1958: $27.18 million
Annual amount 1977: $18.75 million

126

All these figures need to be interpreted and the actual amounts compared in view of the United States tax system. The weightings have been made on the following bases :

Interest rate : 6.75 per cent;

Rate of inflation : 5.00 per cent;

Utilisation of units : 70 per cent (6.132 hours);

Life of units : 30 years.

For its plant at Hartsville the Tennessee Valley Authority (TVA), as a Federal Agency, is not required to pay ordinary local taxes and is liable only for a very small payment in lieu of tax.

The TVA will nonetheless be required to bear the cost of "softening" the impact, accepted in advance in the process of obtaining authorisation for the plant. The money will be spent in close collaboration with and under the supervision of a coordinating committee consisting of fourteen local mayors; it will be used for education, water supply and drainage, health and administrative services (costs), housing (in part), public transport, vocational training, and police. The total expenditure to be borne in this way by the TVA during the period of construction of the plant has been estimated at approximately $10 million.

For the plants at Pilgrim (Plymouth, Mass.) and Millstone (Waterford, Conn.), the tax revenue earned for the town by the power plants is estimated at $4.0 and $5.8 million respectively, their total revenues being $6.86 and $12.56 million respectively in 1974.

vi) The situation in Japan

In Japan electric power stations are liable to the following taxes :

- Fixed Assets Tax

The municipality c a n f i x the tax rate between 1.4 - 2.1 percent of the cost of construction, within which range the recent trend is to raise the tax rate.

The municipality in which the power plant is situated is understood to receive only one-quarter of this tax, which would at the present time be a sum of about 700 million yen per annum for a unit of 925 MWe (Frs. 14 million or $3 million).

The portion of the tax not paid to the municipality in which the power plant is situated may be used to make subsidy payments to the neighbouring municipalities, but the responsibility for its disbursement lies with the government.

- Tax on fuel collected by the Prefecture of Fujui :

This tax appears to have been introduced for a five-year period and to be 5 per cent of the value of the fuel originally loaded or subsequently reloaded in the reactor of a nuclear power plant.

For a reactor of 925 MWe, operating for 6,000 hours per annum, the amount of the tax might be of the order of 450,000 yen annualy (0.8 yen, or 0.2 cents per kilowatt-hour).

Other local governments which house nuclear power plants are beginning to introduce similar taxes.

- Business Office Tax

Corporations are liable to pay 1.5 per cent of their income, but actually this tax revenue is distributed among prefectural governments which house business officesor establishments including power stations.

- The Real Property Acquisition Tax

- The Special Landholding Tax

- Tax for Promoting the Development of Power Resources

The electric companies are required to pay to the Central Government 85 yen per 1,000 KWh they produce and the central government uses this tax revenue for providing grants to local governments which need to finance infrastructure. The amount paid is 450 yen ($2) per year per KW installed, half of which goes to the municipality that houses the power station and half to the neighbouring municipalities. The grant is given for the period between the start of construction and the start of plant operation, and in this respect this may be regarded as a sort of advance payment. For a 1,000 MWe power plant this grant amounts to $2 million per year.

b) <u>Financial facilities</u>

i) <u>General comment</u>

In many countries the local communities can only receive tax revenue obtained from the establishment of a nuclear plant after it has begun to operate.

They are then usually given the funds needed to provide, in good time, any ancillary equipment that may be needed for the arrival of workers on the plant construction site. These facilities are described below.

128

ii) France

The French Government has agreed to make advances against the taxes which a plant will later have to pay. These advances may cover any annual repayment instalments on loans which local authorities have had to arrange with the national financial organisations existing for that purpose, the Caisse des Dépôts et Consignations and the Caisse d'Aide à l'Equipement des Collectivités Locales; the EDF advances bear interest at an agreed rate. These facilities have been regularised by including them in the regulations governing major construction sites.

iii) Belgium

In Belgium finance has been granted to the commune of Tihange against the future tax revenue of the power plant by the company constructing the plant.

iv) United States

For the Skagit plant in the United States, finance has been granted to the local communities against the future tax revenue from the plant. These advances should make it unnecessary for the communities (or other public bodies) to increase their taxes solely on account of the opening of the construction site, as the advances should cover :

- all the additional expenditure involved in taking into schools the children of workers engaged in the construction or operation of the plant;

- all the additional expenditure that may be necessary for public safety.

The conditions of application have been laid down in a very comprehensive and concise contract, and it is the signature of that contract which has made it possible for the land to be occupied by the plant to be classified as an industrial area. This was a precedent that attracted much attention in the United States.

v) Italy

In Italy, the ENEL (nationalised electricity organisation) makes special contributions to the commune in which a nuclear power plant is established designed to cover the expenses of so-called "secondary" town-planning work made necessary by the siting of the plant, excluding what are known as "primary" works, i.e. those serving only the power plant which are carried out by the ENEL at its own expense *.

_____

* The town-planning works classed as "primary" and "secondary" in Italy roughly correspond to the ancillary equipment known in France as "specific" and "preliminary" respectively.

The contribution is paid to the commune as payments by the latter become due for the carrying out of secondary town-planning work, usually before the construction site has been started.

The amount of the special contribution was fixed for 1976 at L.2,200 per KW of installed power and can be revised in accordance with changes in the economic situation *.

vi) Japan

In Japan, the major tax paid by a nuclear power station to the central government is the Fixed Assets Tax which is calculated as 1.4-2.1 per cent on the cost of construction. For a 925 MWe station this investment is about 159.000 million yen (US$530 million), hence the tax is about US $12-13 million. The community which houses the station is given about 25 per cent of that sum, i.e. US$3 million, the rest being distributed through central government. There are, however, other taxes such as the Business Office Tax, the Real Property Acquisition Tax, the Special Landholding Tax, ahd the Tax for promoting the development of power resources.

c)   Special advantages

i)   In Switzerland

In Switzerland, the companies constructing nuclear power plants have given certain benefits to the communes in which the plants are established, as described below :

Goesgen-Daeniken plant

-   Contribution to the construction of a reservoir to supply water through a distribution network to three communes in whose territory the plant is situated, as a fire-fighting reserve;

-   Possible construction of a public footpath in the neighbourhood of the plant;

-   Repair of local roads damaged by vehicles from the plant construction site.

The cost of these benefits will be of the order of Sw.Frs.10 million.

_____

    *   This contribution is a little less than $2 per KW, or $1.8 million for a unit of 925 MWe, 0.25 per cent of the construction price.

Leibstadt plant

- Architectural improvements to the appearance of the plant by lowering the main level of the plant itself by 20 metres and that of the cooling tower by 30 metres.

- Contribution to the administrative costs of the commune :
  while under construction : Sw.Frs.135,000 per annum;
  when in operation : Sw.Frs.250,000 per annum.

- Financial participation in a road improvement programme :
  communal network : Sw.Frs.0.5 million out of a total expenditure of Sw.Frs.2.3 million;
  cantonal network : Sw.Frs.1.0 million out of a total expenditure of Sw.Frs.6.0 million.

- Free supply of 3 million KWh of electric power annually at the gates of the plant (i.e. excluding transport and distribution costs).

Kaiseraugst plant

- Participation in a programme of improvement of communal and cantonal road networks.

- Construction of a public water reservoir to be incorporated in the communal water supply network.

- Construction of a footpath and cycle path between Kaiseraugst and Rheinfelden (for the safety of children going to schools at Rheinfelden, by making it unnecessary for them to use the main roads on which traffic will have become heavier).

The financial contribution of the company constructing the plant is approximately Sw.Frs.8.5 million.

It is considered in Switzerland that during the first years of existence of the plant, the tax revenue and supplementary special benefits accruing to the commune and the Canton from the establishment of a nuclear plant on the terms now planned or agreed will only cover the charges made necessary by that establishment , and it is only later that the plant will begin to provide a very appreciable improvement in the budget situation.

ii) In the United States

Clinton plant

The plant is to be cooled by water taken from and returned to a large cooling pool to be built; it is planned to landscape the pool and its surroundings so as to make a very large recreation area.

The pool will have an area of a little less than 2,000 ha.

This work will make it possible to enjoy walking, boating, swimming, camping, picnics and fishing from the banks.

The number of visiters is expected to be approximately 700,000 each year.

The cost of the work to be borne by the plant construction company has not been given but will be comparatively high, though the local authorities will help in moving roads and bridges.

Pilgrim plant

The pilgrim plant is to be built almost entirely by commuters, and the costs to be borne by the city of Plymouth are accordingly much smaller with the result that the city has been able, with the help of the tax revenue from the plant, to reduce the level of its taxes considerably, the rate being reduced from 42 to 31 between 1965 and 1974.

This reduction in taxes led to a rapid increase in the population (15,400 inhabitants in 1965 to 28,000 in 1975).

iii) In Japan

A special situation has been created in Japan by the fact that a strip of water from the seashore to a line where the depth of water is 12 metres is attributed to the fishing co-operatives who are virtually its owners. In June 1974, for example, a plant which might comprise eight 1,000 MWe units and discharge 40 tonnes of hot water into the sea each day had to pay the fishing industries and fishermen as the price of their acceptance :

- for the purchase of fishing rights in the areas affected by the thermal pollution, 4 billion yen;

- for their free co-operation, 1.45 billion yen;

- for costs to the fishing industry, 170 million yen;

- for indemnities to the families of 20 fishermen who had to abandon this occupation, 75 million yen.

It is with a view to limiting these local demands that laws have been adopted covering the grant of subsidies to local authorities which have already been referred to above; these subsidies will enable local authorities in which a nuclear power plant is situated, and neighbouring authorities, to carry out improvements of value to agriculture, forestry and fisheries, and may also be applied to the building of schools, health establishments, recreational areas and other local amenities.

The compensation represented by such town-planning improvements here has the same aims in view as those of the improvements carried out in France in the same circumstances.

d) Local economic development

i) General conditions

The establishment of a nuclear power plant is usually accompanied by a quite significant development of the local economy. This development is all the greater in the climate of improved budgetary possibilities created by the new tax revenue brought by the power plant, and the collection of taxes of which no mention has so far been made, whether these derive from the operation of the plant, the workers on the site, the increased activity of local traders and craftsmen, the creation of new jobs or, in general, a beneficial change in the social climate.

ii) In France

At St. Laurent des Eaux in France, for instance, after the establishment of the power plant, the population increased both at St. Laurent itself and in the neighbouring communes of La Ferté St. Cyr, Nouan, Muides, Mer, Travers and Beaugency. At St. Laurent the population had fallen to 950 in 1963 compared with 1,600 in 1850; but by 1972 it had risen again to 1,750, the increase being far greater than that resulting from the arrival of EDF staff and their families.

The following were built :

St. Laurent	63 houses or flats
	66 single rooms
	1 firm's restaurant
Mer	232 dwellings
Beaugency	43 dwellings

in all 425 dwellings.

There was a large and continuing increase in the number of applications for building permits at St. Laurent : 27 in 1972 compared to 6 in 1962 before the plant was built.

For the first two units of 500 MWe, the investment was of the order of Frs.1.5 billion and the wage-bill of the site workers was of the order of Frs.300 million at 1974 values; nearly 40 per cent of that sum went to benefit local trade by a turnover of a similar amount.

The personnel operating the plant, numbering 416, were partly recruited on the spot (66, or 17,6 per cent of the total); the corresponding wage-bill was of the order of Frs.12 million and a large part of this second wage-bill was also spent to the benefit of local traders.

It has been estimated on reasonable grounds that the operating of the St. Laurent des Eaux plant, with its first two units of 500 MWe, is now making a contribution to the local economy (commune and regional) of the order of Frs.35 million annually. No traders have gone out of business and local trading is expanding strongly.

The area of cultivable land occupied by the plant and lost to local farmers has been comparatively small compared with the total cultivated area in the commune, and the number of farms has not fallen.

At Fessenheim, a very careful enquiry made by the EDF in 1974 showed that the considerable investment represented by the construction of the first 500 MWe unit of the nuclear plant, nearly Frs.2 billion at that time, had brought about considerable economic expansion locally.

Local trade had reached a turnover of the order of 75 per cent of the wage-bill of the construction site workers, or nearly Frs.100 million; and traders in the neighbourhood of the plant had been able to make large investments much earlier than would have been possible if the plant had not been constructed.

Local firms had received orders estimated as follows :

–	First under contract to EDF (excluding Alsthom)	Frs.47.5 million
–	Sub-contractors	Frs.15.2 million
–	Firms building EDF housing	Frs.50.0 million
	Total	Frs.112.7 million

This total does not include the sum paid to Alsthom, which is a special case among the local firms at Fessenheim (its local plant being at Belfort); the order placed with Alsthom was for Frs.280 million.

Even without taking this large contract into account, the participation by local firms in supplies and works has been very large despite the requirement for the EDF not to confine its invitations to tender to those firms alone; the same was true in general of the nuclear plants at Chinon, Chooz, Brennylys and St. Laurent des Eaux as well as for the traditional thermal plants.

In the same way, local firms will have an opportunity to continue doing business when the Fessenheim plant is in operation providing supplies and services needed by the operating staff or the plant itself, which may result in permanent employment for 50 or 60 people.

Local trade should also benefit from the visits of EDF travelling teams or specialised firms coming to reinforce the local EDF personnel for periodic heavy maintenance or repair work. These intermittent visits will be roughly equivalent to the permanent presence of some 30 people.

Lastly, local trade is benefiting from the special kind of tourism consisting in large numbers of visitors to the plant (more than 10,000 annually at present).

iii) In Belgium

At Tihange and Doel in Belgium the prospect of local economic expansion was an important factor in favour of accepting the nuclear plant at Tihange in particular, so far as the creation of new jobs was concerned.

With the help of the financial advantages which the commune of Tihange received it was able to buy land and develop it for industrial and residential property, thus promoting the creation of permanent new employment.

iv) In Switzerland

In Switzerland, as elsewhere, the setting up of a nuclear power plant is expected to lead to significant local economic development :

- the creation of 50 to 80 new jobs;
- an expansion of local trade benefiting from purchases made by the workers on the construction site and the operating staff of the plant;
- the placing of orders with local firms and artisans wherever possible, in accordance with the promise usually made by the constructors of the plant.

v) In the United States

For the Skagit nuclear plant in the United States,
the local population's acceptance was in the end mainly
motivated by expectations of benefiting from the economic expansion which should result from the construction of the plant (and with no fear in any increase
in local taxes thanks to the financial advances made
to local communities, as described in 3(b)iv above).
The expected economic development was as follows :

- a general expansion induced by a comfortable
budget situation resulting from the additional
tax revenue yielded by operating the plant,
throughout its probable life of 30 years, increases being of the order of 21 per cent of
total revenue and 25 per cent on property tax
in the County;

- offers of jobs to young people coming on to
the labour market, throughout the duration of
the construction work, resulting directly or
indirectly from the opening of the building
site.

4.   Settling the Details of Compensation Guarantee
of Payment.

a)   General conditions

The process of studying and authorising the plans
for a nuclear power plant usually provides all the information needed to decide whether compensation should
be paid for the disadvantages of proximity in the
particular case of that plant, and to determine how
such compensation shall be paid; this process also
leaves open the possibility of guaranteeing to the
local population that the arrangements for compensation
will be put into effect as planned if the plant is
constructed.

Confirmation of this will be found below in a comparatively detailed description of the procedure adopted in France, which corresponds fairly closely in its
main features to what is now done in most countries,
subject to adaptation to special legal or administrative provisions of which we shall give a few
examples.

b)   Procedure adopted in France

The procedure for studying and authorising plans
adopted in France has two main features :

i) Study of the project is accompanied by an information campaign addressed to the local population, and in general agreement with it, at
a very early stage when it is no more than an
outline.

ii) The administrative authorisation procedure
includes a public interest statement (Déclaration d'Utilité Publique - DUP) leading to a
very open public enquiry.

The preliminary choice of the site not only has
to meet technical criteria (supply of cooling water,
nature and possible classification of the soil, industrial environment, length of transmission lines to
the general supply network, road and rail access, etc.)
but also aesthetic and ecological criteria reflecting
the growing interest in the protection of the environment and the improvement of the quality of life (disturbance of the landscape, heat discharge into watercourses or the sea, discharge of steam into the atmosphere by cooling towers, if any); and lastly, economic and human criteria (such as those of local economic development). A large amount of information is
assembled about the technical criteria by the EDF
engineers and specialists in various disciplines, and
about the aesthetic, ecological, economic and human
criteria mainly by the local population.

The reactions of the local population are sought
at a very early stage : in 1974 for example, the
Minister concerned asked the Préfets to obtain the
views of the departmental and regional Conseils géné-
raux on 37 sites, although they were only prima facie
likely to be suitable for a nuclear power plant.

"The projects submitted to the public are usually
at an early stage and have not been worked out
in great detail; while the analyses done at this
stage at the level of communes can result in the
proposed siting being slightly modified to take
account of the use being made of the land or to
fix means of access in the light of local needs.
At that time preliminary exchanges of views are
held to enable the Mayor to assess as precisely
as possible the foreseeable economic impacts of
the construction. It is thus on the basis of sound
information that the Municipal Council and the
Mayor can decide on their position, make it known
to the Préfet and the electors and take part in
the Government's decision to abandon the project
or go ahead with it by authorising the EDF to
apply for a "Public Interest Statement." *

During the period referred to above everthing
necessary has been done to enable a useful discussion
to be held about reducing the disadvantages of proximity or paying compensation for them.

---

* Extract from the handbook "Une centrale nucléaire dans
la commune" published by the Ministry for Industry and Research,
Délégation générale à l'Energie.

The administrative procedures for authorisation include general law procedures and specifically nuclear ones.

The general law procedures provide for the traditional statutory enquiries concerning, in particular, authorisation to take water from, and return it to, a watercourse or the sea.

The public enquiries provided for in the regulations governing authorisation of the setting up of the plant, as a basic nuclear establishment, may be confused with the enquiry prior to issuing the Public Interest Statement, which may also cover establishments classified as dangerous, unhealthy or unpleasant established within the perimeter of the nuclear plant.

The Public Interest Statement is not a legal requirement in France except in cases where expropriations have to be made; for all electric power plants, however, whether of the traditional thermal kind or nuclear, it is no longer the consideration of expropriation which matters but the fact that the DUP provides official recognition by the Conseil d'Etat that the proposed operation is desirable.

The enquiry held before the issue of the DUP is, because of its extremely wide scope, an especially suitable means of giving the public final information and obtaining its final agreement as well as that of the institutions responsible for defending the public interest; this is especially true as regards the socio-economic impact and balance of the future plant.

The application for the DUP is made by the EDF on completion of its study of the project and must be accompanied by a very comprehensive supporting dossier comprising :

- an explanatory note indicating, in particular, the purpose of the operation;

- the site plan;

- the general plan of work;

- the main features of the most important buildings;

- a short estimate of expenditure;

- a note on the safety of the nuclear installations;

- a note on any establishments classified as dangerous, unhealthy or unpleasant situated within the perimeter of the plant;

- an architectural study;

138

- a study of the impact on the environment, with a short reference list, a description of studies undertaken, a preliminary statement of the events connected with the establishment, a note on how it will fit into its surroundings, especially from the architectural point of view, and a report on measures to be taken to avoid damage to the environment and nuisances;

- a note on those benefits expected from the operation which appear sufficiently convincing, despite possible inconveniences, to justify a declaration that the plant is in the public interest;

- a note on the reasons for selecting the project submitted among the other possibilities, and, where appropriate, for preferring it to other projects prepared outside the Administration.

The widest possible publicity has to be given to the making of the application for a DUP and a copy of the application and the attached dossier is addressed by the Préfet to the Département, the elected representatives, important people and organisations concerned, members of Parliament, regional councillors, members of regional economic and social committees, general departmental councillors, mayors, chambers of commerce and industry, chambers of agriculture and trades, etc.

The decision to open the enquiry leading to the DUP is taken by the Inter-departmental Office of Industry and Mining (Service interdépartemental de l'Industrie et des Mines) after consultation with the various administrative services involved and, where appropriate, the preparation of an enquiry dossier based on that submitted by the EDF in support of its application. It is at the stage of preparing this dossier that details of the compensation arrangements may be introduced.

The DUP enquiry must be held at the Prefecture of the Départment and the Mairie of the Commune in which the plant is to be established, and, where appropriate, in the Mairies of communes having a part of their territory within 5 kilometres of the plant, and in the Préfecture of those communes.

The enquiry must remain open for at least a fortnight, but this period may be extended to six weeks or even two months for such important and delicate operations as the setting up of a nuclear plant usually involves.

The DUP enquiry has to be very open and very comprehensive; it must make it possible for the population concerned to be well-informed, and for concerted view to develop broadly and freely; applications for compensation may be made on any points which have not already been covered in the enquiry dossier, and these then acquire an official status; all the statements made in the course of the enquiry must be in writing and must appear in the report of the Committee of Enquiry either as transcriptions or by appending an original note.

After submission to the Préfet or Préfets, the enquiry dossier is given very careful consideration in an interministerial conference following which, if the Government decides in the light of the results of the enquiry to go ahead with the operation, a draft decree embodying the DUP is prepared and placed before the Conseil d'Etat by the Minister concerned, accompanied by the enquiry dossier.

If the Conseil d'Etat decides in its favour, the DUP decree is adopted by the Government, taking account of any proposals made by the Conseil d'Etat.

The issuance of the DUP thus constitutes recognition by the highest responsible body, fully independent of the Government, that the establishment of the plant is desirable.

It will be seen, without the need for us to stress the point, that the examination by the Conseil d'Etat covers the question of whether any compensation measures which have been asked for by the local population or by the Administration, up to the final stage of the administrative procedure, are in the public interest. This procedure provides a guarantee for the local population , if it were necessary, that the measures judged by the Conseil d'Etat to be of value, clearly defined in the documents referred to in the DUP, and appended to that declaration, will be carried out.

c) Procedure in Belgium

The local public enquiry opened under the specifically nuclear procedure remains open for only a fortnight.

The result of that initial enquiry is however followed up in turn by : an examination and opinion of the municipal council; an examination and opinion of the permanent delegation of the Province; and an examination and opinion of a special administrative commission, which may co-opt Belgian and foreign specialists and consult Euratom.

In all, this procedure may require several years for each plant.

The final decision to grant authorisation is taken by a Royal Order on the recommendation of the special commission, though with no guarantee that it will not be refused even in those circumstances.

In practice compensation for the disadvantages of proximity will probably be asked for not only during the initial local enquiry but also during the subsequent proceedings, or directly by those concerned, though in the latter case this will be through the local elected representatives.

d) Procedure in Switzerland

The administrative procedure for authorising nuclear power plants is still governed in Switzerland by the provisions of the organic law of 1969 which is a regulatory measure giving the local population only very limited possibilities of participating in the drawing up of authorisations.

In practice this legal situation has not prevented local communities, wherever a plant has been established from enjoying compensation peasures of the same kind and on the same scale as in France, for example, where the regulations have evolved further.

There is however general agreement that the organic law of 1969 is no longer suited to the present situation and must be amended.

While it is proposed to make an order which would allow the population concerned to take part in the near future in the procedure for authorising nuclear power plants by enabling each citizen to defend his or her interests at law, and also to be sure that no plant will be built which does not meet a Federal and local need or in a word that no one will be needlessly exposed to the disadvantages of proximity of a nuclear power plant.

It is hoped with the help of this order to be able to avoid a referendum under which construction of any nuclear power plant would be subject to its acceptance by the commune and Canton in which it was situated, and by neighbouring communes and Cantons.

e) Procedure in the United States

Public enquiries in the United States take the form of hearings at which arguments may be presented or supported orally.

The enquiry dossier prepared by the authorities constitutes what is called the "environmental impact statement".

This statement reflects the highly developed and detailed studies made by the administrative offices concerned and the discussions held among them in order to limit the impact, and particularly the disadvantages of proximity or ensure that compensation is paid for them.

When additional measures are proposed for this purpose at the hearings they are recorded in the minutes of the meetings, which are appended to the environmental statement. These documents are referred to in the decision authorising the construction of the plant, and through this alone the local population can be sure that all the planned provisions to limit the disadvantages of proximity or to pay compensation for it, will be effectively implemented. This guarantee can also be given in the form of contracts (as was done in the case of the Skagit and Hatsville plants, referred to above).

OECD SALES AGENTS
DÉPOSITAIRES DES PUBLICATIONS DE L'OCDE

ARGENTINA – ARGENTINE
Carlos Hirsch S.R.L., Florida 165, 4° Piso (Galería Guemes)
1333 BUENOS AIRES, Tel. 33-1787-2391 Y 30-7122

AUSTRALIA – AUSTRALIE
Australia & New Zealand Book Company Pty Ltd.,
23 Cross Street, (P.O.B. 459)
BROOKVALE NSW 2100   Tel. 938-2244

AUSTRIA – AUTRICHE
OECD Publications and Information Center
4 Simrockstrasse 5300 BONN   Tel. (0228) 21 60 45
Local Agent:
Gerold and Co., Graben 31, WIEN I. Tel. 52.22.35

BELGIUM – BELGIQUE
LCLS
44 rue Otlet. B 1070 BRUXELLES . Tel. 02-521 28 13

BRAZIL – BRÉSIL
Mestre Jou S.A., Rua Guaipá 518,
Caixa Postal 24090, 05089 SAO PAULO 10. Tel. 261-1920
Rua Senador Dantas 19 s/205-6, RIO DE JANEIRO GB.
Tel. 232-07. 32

CANADA
Renouf Publishing Company Limited,
2182 St. Catherine Street West,
MONTREAL, Quebec H3H 1M7 Tel. (514) 937-3519

DENMARK – DANEMARK
Munksgaards Boghandel,
Nørregade 6, 1165 KØBENHAVN K. Tel. (01) 12 85 70

FINLAND – FINLANDE
Akateeminen Kirjakauppa
Keskuskatu 1, 00100 HELSINKI 10. Tel. 65-11-22

FRANCE
Bureau des Publications de l'OCDE,
2 rue André-Pascal, 75775 PARIS CEDEX 16. Tel. (1) 524.81.67
Principal correspondant :
13602 AIX-EN-PROVENCE : Librairie de l'Université.
Tel. 26.18.08

GERMANY – ALLEMAGNE
OECD Publications and Information Center
4 Simrockstrasse 5300 BONN   Tel. (0228) 21 60 45

GREECE – GRÈCE
Librairie Kauffmann, 28 rue du Stade,
ATHÈNES 132. Tel. 322.21.60

HONG-KONG
Government Information Services,
Sales and Publications Office, Baskerville House, 2nd floor,
13 Duddell Street, Central.   Tel. 5-214375

ICELAND – ISLANDE
Snaebjörn Jónsson and Co., h.f.,
Hafnarstraeti 4 and 9, P.O.B. 1131, REYKJAVIK.
Tel. 13133/14281/11936

INDIA – INDE
Oxford Book and Stationery Co.:
NEW DELHI, Scindia House. Tel. 45896
CALCUTTA, 17 Park Street. Tel. 240832

INDONESIA – INDONÉSIE
PDIN-LIPI, P.O. Box 3065/JKT., JAKARTA, Tel. 583467

IRELAND – IRLANDE
TDC Publishers — Library Suppliers
12 North Frederick Street, Dublin 1   Tel. 744835-749677

ITALY – ITALIE
Libreria Commissionaria Sansoni:
Via Lamarmora 45, 50121 FIRENZE. Tel. 579751
Via Bartolini 29, 20155 MILANO. Tel. 365083
Sub-depositari:
Editrice e Libreria Herder,
Piazza Montecitorio 120, 00 186 ROMA. Tel. 6794628
Libreria Hoepli, Via Hoepli 5, 20121 MILANO. Tel. 865446
Libreria Lattes, Via Garibaldi 3, 10122 TORINO. Tel. 519274
La diffusione delle edizioni OCSE è inoltre assicurata dalle migliori
librerie nelle città più importanti.

JAPAN – JAPON
OECD Publications and Information Center,
Landic Akasaka Bldg., 2-3-4 Akasaka,
Minato-ku, TOKYO 107 Tel. 586-2016

KOREA - CORÉE
Pan Korea Book Corporation,
P.O.Box n° 101 Kwangwhamun, SÉOUL. Tel. 72-7369

LEBANON – LIBAN
Documenta Scientifica/Redico,
Edison Building, Bliss Street, P.O.Box 5641, BEIRUT.
Tel. 354429–344425

MALAYSIA – MALAISIE
and/et SINGAPORE-SINGAPOUR
University of Malaya Co-operative Bookshop Ltd.
P.O. Box 1127, Jalan Pantai Baru
KUALA LUMPUR Tel. 51425, 54058, 54361

THE NETHERLANDS – PAYS-BAS
Staatsuitgeverij
Verzendboekhandel Chr. Plantijnstraat
S-GRAVENHAGE   Tel. nr. 070-789911
Voor bestellingen: Tel. 070-789208

NEW ZEALAND – NOUVELLE-ZELANDE
The Publications Manager,
Government Printing Office,
WELLINGTON: Mulgrave Street (Private Bag).
World Trade Centre, Cubacade, Cuba Street,
Rutherford House, Lambton Quay, Tel. 737-320
AUCKLAND: Rutland Street (P.O.Box 5344), Tel. 32.919
CHRISTCHURCH: 130 Oxford Tce (Private Bag), Tel. 50.331
HAMILTON: Barton Street (P.O.Box 857), Tel. 80.103
DUNEDIN: T & G Building, Princes Street (P.O.Box 1104),
Tel. 78.294

NORWAY – NORVÈGE
J.G. TANUM A/S Karl Johansgate 43
P.O. Box 1177 Sentrum OSLO 1   Tel (02) 80 12 60

PAKISTAN
Mirza Book Agency, 65 Shahrah Quaid-E-Azam, LAHORE 3.
Tel. 66839

PHILIPPINES
National Book Store, Inc.
Library Services Division, P.O.Box 1934, Manila,
Tel. Nos. 49-43-06 to 09 40-53-45 49-45-12

PORTUGAL
Livraria Portugal, Rua do Carmo 70-74,
1117 LISBOA CODEX. Tel. 360582/3

SPAIN – ESPAGNE
Mundi-Prensa Libros, S.A.
Castelló 37, Apartado 1223, MADRID-1. Tel. 275.46.55
Libreria Bastinos, Pelayo, 52, BARCELONA 1. Tel. 222.06.00

SWEDEN – SUÈDE
AB CE Fritzes Kungl Hovbokhandel,
Box 16 356, S 103 27 STH, Regeringsgatan 12,
DS STOCKHOLM. Tel. 08/23 89 00

SWITZERLAND – SUISSE
OECD Publications and Information Center
4 Simrockstrasse 5300 BONN Tel. (0228) 21 60 45
Agents locaux :
Librairie Payot, 6 rue Grenus, 1211 GENÈVE 11. Tel: 022.31.89.50
Freihofer A.G., Weinbergstr. 109, CH-8006 Zürich Tel: 01-3624282

TAIWAN – FORMOSE
National Book Company,
84-5 Sing Sung South Rd., Sec. 3, TAIPEI 107. Tel. 321.0698

THAILAND – THAILANDE
Suksit Siam Co., Ltd., 1715 Rama IV Rd.
Samyan, BANGKOK 5   Tel. 2511630

UNITED KINGDOM – ROYAUME-UNI
H.M. Stationery Office, P.O.B. 569,
LONDON SE1 9 NH. Tel. 01-928-6977, Ext. 410 or
49 High Holborn, LONDON WC1V 6 HB (personal callers)
Branches at: EDINBURGH, BIRMINGHAM, BRISTOL,
MANCHESTER, CARDIFF, BELFAST.

UNITED STATES OF AMERICA – ÉTATS-UNIS
OECD Publications and Information Center, Suite 1207,
1750 Pennsylvania Ave., N.W. WASHINGTON, D.C. 20006.
Tel. (202)724 1857

VENEZUELA
Libreria del Este, Avda. F. Miranda 52, Edificio Galipán,
CARACAS 106. Tel. 32 23 01/33 26 04/33 24 73

YUGOSLAVIA – YOUGOSLAVIE
Jugoslovenska Knjiga, Terazije 27, P.O.B. 36, BEOGRAD.
Tel. 621-992

Les commandes provenant de pays où l'OCDE n'a pas encore désigné de dépositaire peuvent être adressées à :
OCDE, Bureau des Publications, 2 rue André-Pascal, 75775 PARIS CEDEX 16.
Orders and inquiries from countries where sales agents have not yet been appointed may be sent to:
OECD, Publications Office, 2 rue André-Pascal, 75775 PARIS CEDEX 16.

8-1980

OECD PUBLICATIONS, 2 rue André-Pascal, 75775 Paris Cedex 16 - No. 41 265   1980
PRINTED IN FRANCE
(800 TD 97 79 07 1) ISBN 92-64-11986-8